American Slavery, Atlantic Slavery, and Beyond

U.S. History in International Perspective

Peter N. Stearns and Thomas W. Zeiler, series editors

Now Available

Revolutions in Sorrow: The American Experience of Death in Global Perspective, by Peter N. Stearns
From Alienation to Addiction: Modern American Work in Global Historical Perspective, by Peter N. Stearns
Diverse Nations: Explorations in the History of Racial & Ethnic Pluralism, by George M. Fredrickson
American Slavery, Atlantic Slavery, and Beyond: The U.S. "Peculiar Institution" in International Perspective, by Enrico Dal Lago

Forthcoming

American Trade in International Perspective: 1890s to the Present, by Francine McKenzie
The Global Great Depression and the Coming of WWII, by John E. Moser
Indians and Invaders: North American Native Peoples in International Perspective, by Roger L. Nichols

American Slavery, Atlantic Slavery, and Beyond

The U.S. "Peculiar Institution" in International Perspective

Enrico Dal Lago

Paradigm Publishers
Boulder • London

Published in the United States by Paradigm Publishers, 5589 Arapahoe Avenue, Boulder, CO 80303 USA.

Paradigm Publishers is the trade name of Birkenkamp & Company, LLC, Dean Birkenkamp, President and Publisher.

Library of Congress Cataloging-in-Publication Data

Dal Lago, Enrico, 1966–
American slavery, Atlantic slavery, and beyond : the U.S. "peculiar institution" in international perspective / Enrico Dal Lago.
p. cm. — (U.S. history in international perspective)
Includes bibliographical references and index.
ISBN 978-1-59451-585-9 (paperback : alk. paper)
1. Slavery—United States—History. 2. Slavery—Economic aspects—United States—History. 3. Slavery—Atlantic Ocean Region—History. 4. Slavery—Economic aspects—Atlantic Ocean Region—History. 5. Cotton growing—Economic aspects—Southern States—History. 6. International economic relations—History—19th century. 7. United States—History—Civil War, 1861–1865—Causes. I. Title.
E441.D18 2012
306.3'620973—dc23

2011042175

Printed and bound in the United States of America on acid free paper that meets the standards of the American National Standard for Permanence of Paper for Printed Library Materials.

Designed and Typeset by Straight Creek Bookmakers.

17 16 15 14 13 1 2 3 4 5

CONTENTS

LIST OF MAPS

PREFACE

This book's aim is to provide an interpretive summary of past and present scholarship focused on comparative history of American slavery and to suggest future directions of inquiry. This research field is particularly large and complex, but still full of possibilities for new openings. Some of these openings are presented here as possible avenues for future comparative studies, but there are many more avenues that I have not indicated in my study. In both cases, I hope the suggestions presented here will provide a stimulus to undertake the difficult task of engaging with comparative historical studies of the U.S. "peculiar institution"—a particularly appropriate expression, since some prominent nineteenth-century southerners used it to describe the specificity of American slavery in comparison with other labor systems.[1]

The introduction to this book provides a general overview of the development of comparative historical studies, with particular reference to the U.S. South and its Euro-American context, from the 1940s to the present. The last pages of the introduction cover the current debates on the relationship between transnational and comparative approaches to the American past.

The book is divided into two parts:

Part A: American Slavery in the Atlantic World focuses on the continental context of colonial, revolutionary, and nineteenth-century American slavery in comparison with New World slave societies, particularly Brazil and the Caribbean. Part A also pays particular attention to American slavery's Atlantic context—in reference to both Europe and Africa—with a specific focus on the slaveholding elites' ideologies and the lives of the slaves. Both are fields of study in which the comparative and transnational approaches have yielded particularly important insights.

Part B: American Slavery in the Euro-American World is characterized by a wider comparative approach, focusing on Euro-American perspectives on American slavery. Specifically, comparison in Part B focuses on slavery in the nineteenth-century U.S. South, and both contemporary and different forms of unfree and nominally free labor, especially eastern European serfdom and Mediterranean sharecropping and tenancy. Part B also places American abolitionism and the American Civil War within the wider context of Euro-American nationalist movements. In all these cases, the comparative approach has yielded particularly valuable insights, especially on the definitions of freedom and unfreedom and their different meanings in America and Europe.

Ultimately, this book intends to argue that, while focus on the Atlantic context has given us invaluable contributions on the history, development, and characteristics of American slavery, it is now time to acknowledge the equally crucial importance of the wider Euro-American context for a correct understanding not just of the U.S. "peculiar institution," but also of American abolitionism and the American Civil War and emancipation within the global nineteenth century of which they were part. In fact, as C. A. Bayly has shown in *The Birth of the Modern World* (2004), the nineteenth century was truly modern also for the global repercussions of its major economic, social, and political features and events.[2]

ACKNOWLEDGMENTS

I wish to express my sincere thanks to all the institutions and people that have helped me in writing my book. First and foremost, I wish to thank the National University of Ireland, Galway's "Millennium Fund," under whose project "Slave Systems" I have conducted part of the necessary research, especially at the Boston Public Library.

I also wish to thank the staff at the Interlibrary Loan services at the James Hardiman Library at NUI, Galway, for their invaluable help, and also, for their support, my Head of Discipline, Professor Steven Ellis, and my colleagues at NUI, Galway's History Department, School of Humanities; the former director of the Moore Institute, Professor Nicholas Canny, from whom I have learned a great deal about the Atlantic world over the course of the years; and also the students in the final-year seminar course on "Slavery and Emancipation," the past students in the MA in History, to whom I hope I have managed to teach the basics of comparative history and comparative slavery, and my Ph.D. students, all engaged in fascinating comparative and/or transnational topics.

Earlier and somewhat different versions of parts of the introduction and Chapters Three and Four were published in two articles: "Second Slavery, Second Serfdom, and Beyond: The Atlantic Plantation System and the Eastern and Southern European Landed Estate System in Comparative Perspective, 1800–1860," *Review* (Fernand Braudel Center) 32:4 (2009), 391-420; and "Comparative Slavery" in Robert L. Paquette and Mark M. Smith, eds., *The Oxford Handbook of Slavery in the Americas* (New York: Oxford University Press, 2010), 664-684.

Also, I have presented three papers based on shorter versions of other parts of the introduction and of two other chapters of the book: "The New Frontiers of Comparative History: From the 'Transnational Turn' to World History, and Beyond," keynote lecture in the International Workshop "Comparative History in a Transnational Age," University of Warwick, UK, June 2009; "Slavery, Antislavery, and Nationalism on the Two Sides of the Atlantic: In Search of Forgotten Links between the Nineteenth-Century United States and Europe," presented at the International Conference "The Nineteenth Century and the New Frontiers of Slavery and Freedom," *Universidade Federal do Estado do Rio De Janeiro* and *Universidade Severino Sombra,* Vassouras, Brazil, August 2009; and "The American Civil War, Emancipation, and Nation-Building: A Comparative Perspective," presented at the International Conference on "The Politics of the Second Slavery," Fernand Braudel Center, Binghamton University, NY, October 2010.

A number of friends and colleagues have heard, read, or commented on the above papers and articles, offering crucial insights that have helped me to understand and conceptualize better my study. I wish to thank, first and foremost, the friends of the "second slavery" network, based in the Fernand Braudel Center, Binghamton University, especially Dale Tomich, Rafael de Bivar Marquese, Robin Blackburn, Ricardo Salles, Ed Baptist, Edward Rugemer, Wazir Mohammed, Luiz Felipe de Alencastro, Jason Moore, and Christopher Schmidt-Nowara. I also wish to thank for their important input, in different forms, Peter Stearns, Peter Kolchin, Mark Smith, Catherine Clinton, Philip Morgan, Bruce Levine, Kim LoPrete, Gearoid Barry, Dan Carey, Norwood Andrews, Niall Whelehan, Adrian Paterson, Patrick and Anne Towers, and Stefano Dal Lago. I wish to give special thanks to Madeline, for her wonderful friendship and constant support. Finally, I wish to thank Leslie Singer Lomas, Jennifer Knerr, Terra Ann Dunham, and Candace English at Paradigm Publishers for all their help and understanding, and Sinead Armstrong for having again produced beautiful maps that reflect what I had in mind. My biggest thanks, though, go to my parents, Olinto and Rosa Dal Lago, for being extremely patient and helpful during the writing of my book.

INTRODUCTION

FROM AMERICAN SLAVERY TO ATLANTIC SLAVERY, AND BEYOND

Historians of both American slavery and slavery in the Americas have long favored comparison as a method of investigation. Historical monographs have compared aspects of different slave societies in the New World, especially the U.S. South, for more than 60 years. More general studies have investigated the rise and development of unfree labor in the Americas and elsewhere, and have contributed significantly to broadening our perspectives.

Historical comparison has shed new light on controversial issues and, more importantly, it has generated new interpretations on topics such as the following:

- The emergence of unfree labor systems
- The relation between modernity and free and unfree agrarian labor
- The coexistence of capitalist and precapitalist features in the elites' ideologies and behaviors
- The uniqueness of modern abolitionist movements and of processes of emancipation in the Americas

When analyzing the development of the entire comparative historiography on New World slavery, it becomes clear that comparative historians have not followed a single comparative methodology. A number of scholars have relied particularly on seminal methodological articles written by Marc Bloch and by Theda Skocpol and Margaret Somers; others, instead, have preferred to develop their own methodology of comparison.[1]

In this respect, the appearance of Peter Kolchin's *A Sphinx on the American Land* in 2003 has been a landmark in comparative historical studies. Through his book, Kolchin has attempted for the first time to construct a systematic framework for understanding comparative scholarship dealing with the nineteenth-century American South. Essentially, Kolchin has outlined three types of comparison:

- Comparison between the South and the "un-South," or the North
- Comparison between the "many souths," or the different geographical, socioeconomic, and political types of South
- Comparison between the South and the "other souths," or other regions of the world with comparable historical, socioeconomic, and/or political features[2]

To those who study American slavery in comparative perspective, Kolchin's framework offers an opportunity to think carefully and precisely about the continental scale of New World slavery. By taking a step forward, it allows us to characterize the main methodologies that different comparative historians of slavery in the Americas have employed for the past 60 years.

Three main types of comparative analysis appear to have shaped the historiography:

- Comparisons between the slave societies of either specific regions of, or the entirety of, North America on one hand, and Latin America and the Caribbean on the other hand
- Comparisons between slave societies of different regions of the English-speaking Americas on one hand, or between slave societies of different regions of the non-English-speaking Americas on the other hand
- Comparisons between slave societies of either specific regions of, or the entirety of, the New World on one hand, and either slave or unslave labor societies in the Old World on the other hand

Still following Kolchin, we can also fruitfully make a distinction between three types of studies. First, some studies focus on explicit, or *rigorous,* comparisons, in which a sustained comparative study of two or more cases is the object of the research. Other studies contain implicit, or *soft,* comparisons, in which comparative points are offered but remain largely underdeveloped. Finally, comparison can be *synchronic,* when the case studies compared are contemporaneous, or *diachronic,* when the case studies compared belong to different historical epochs.[3]

In this introduction, I will review the historiography of comparative slavery in the Americas and in the wider Euro-American world, in order to identify the main trends and changes it went through. These changes affected deeply the course of comparative studies focusing on slavery in the

U.S. South. This will start familiarizing the reader with the different ways in which comparative perspectives enhance our understanding of the different historical phenomena—chief among them capitalism—historically originated in the Old World and associated with the rise and spread of the Atlantic slave system in the New World.

In the last section of the introduction, I will discuss some considerations regarding the transnational turn in American history and world history, and how it relates to both comparative history and studies on the comparative history of slavery and other systems of labor. I will end the introduction by arguing that it is possible to integrate the search for transnational connections with the comparative study of slave societies and nonslave societies. There is a great deal to learn from a combined research methodology that focuses as much on connections as on investigation of similarities and differences between case studies.

ORIGINS TO EARLY 1970S: AMERICAN SLAVERY AND COMPARATIVE SLAVERY

In general, both the methodology of comparative slavery and the issues on which comparative historians of American slavery currently focus result from the developments of a long historiographical tradition that began with the publication of Frank Tannenbaum's *Slave and Citizen* (1947). With his book, Tannenbaum simultaneously provided the then few comparative historians with a new field of study—slavery in the Americas—and also gave slavery historians a new methodology—comparative history.[4]

Brilliantly applying Bloch's central comparative tenet regarding the detection of a similarity between the facts observed and the dissimilarity between the situations in which they had arisen, Tannenbaum sought to explain the now disproven assumption that slavery in the United States, despite its apparent similarity to slavery in Latin America, appeared much harsher than the latter in legal terms. He found the explanation in the United States' much more exploitative version of capitalism, unmitigated by the Catholic Church. Although not intended as a specifically comparative study, Stanley Elkins's later work, *Slavery* (1959), did much to reinforce Tannenbaum's idea and led to the so-called "Tannenbaum-Elkins hypothesis."[5]

Thus, the first steps in comparative slavery in the Americas, significantly taken by a Latin American specialist and a U.S. specialist, led very early to the idea—represented by comparative studies listed as type A earlier in the introduction—that methodologically, comparisons were best focused on particular macro features, such as law or race, and that the comparisons' objective was to show the reasons for the different characteristics assumed by the U.S. slave system as opposed to its Latin American counterparts.

This main idea informed early comparative works by Herbert Klein and Carl Degler. These scholars used a variant of the method of *juxtaposition,* in which two cases are juxtaposed and then conclusions are drawn from their comparison, to proof or disproof corollaries of the Tannenbaum-Elkins hypothesis. In *Slavery in the Americas* (1967), Herbert Klein focused his comparative study on Virginia and Cuba in the nineteenth century, and in *Neither Black Nor White* (1971), Carl Degler focused his comparison on the United States as a whole and Brazil, taking a broad view at the entire course of their history. While Klein ended up offering further support to Tannenbaum and Elkins, Degler arrived at opposite conclusions. Degler argued, in particular, that the slave system in Brazil was more rigid and brutal than the slave system in the United States, as the much higher mortality rate of Brazilian slaves, due to the harsher working conditions, testified. Degler also provided a model of comparison in regard to the race issue in the two countries and attempted to explain the marked differences between them with a higher degree of racial rigidity in the United States than in Brazil.[6]

In the same period during which Herbert Klein and Carl Degler wrote their pioneering studies, Eugene Genovese elaborated his complex theory on the antebellum U.S. South's slave system. Anticipating future developments, Genovese often added, in support of his hypotheses, important notations and discussions with comparative themes. Even though he never produced a full-fledged monograph based on a particular comparative theme, in the works he published in the 1960s and early 1970s, especially *The World the Slaveholders Made* (1969) and *Roll, Jordan, Roll* (1974), Genovese used examples from his studies of other slave societies of the American continent, particularly Brazil, in order to highlight particular aspects of his notion that the slave system of the antebellum U.S. South was unique. Particularly in *The World the Slaveholders Made,* Genovese compared diachronically New World slave societies with their European antecedents, in order to reinforce the idea that the U.S. slave system was exceptional.[7]

To Genovese, the exceptional and unique character of American slavery lay in the particular prebourgeois features of the ideology and behavior of southern planters, and in the specific characteristics of the type of paternalism that masters exercised on their slaves in southern plantations. According to Genovese, these characteristics made the southern planters' paternalism different not only from the master-slave relationships of other regions of the American continent, but also from the master-bondsmen relationship of other regions of the world.

LATE 1960S TO EARLY 1980S: FROM AMERICAN SLAVERY TO ATLANTIC SLAVERY

By the 1970s, the historiographic revolution centering on *slave agency*—the broadly constituted effort to place slaves themselves at the center of the narrative—reached its heyday. U.S.-originated models of interpretation of the origins and main features of slavery were influencing, and were themselves influenced by, research on slave societies in the Caribbean and Latin America. This cross-fertilization led to a proliferation of comparative points in otherwise circumscribed studies on slave life in different parts of the New World. It also led to a few explicit comparative studies of type B, as listed earlier in the introduction, above all a seminal one by Gwendolyn Midlo Hall, *Social Control in Slave Plantation Societies* (1971). This work focused on slavery and racism as a means to preserve social order in two colonies with large slave populations, such as eighteenth-century St. Domingue and nineteenth-century Cuba.[8]

At the same time, the rereading and rediscovery of older studies by great slavery scholars such as Gilberto Freyre, C. L. R. James, and Eric Williams—which came to be seen as points of departure for comparative research on major themes touching on all slave societies in the New World—helped create the preconditions for a radical change in the overall interpretive framework of slavery in the Americas. Already, Genovese's *From Rebellion to Revolution* (1979) referred to a comprehensive interpretation of slave revolts that went beyond the continental American horizon and looked at broader comparative points. This anticipated later type C comparative studies, as listed earlier in this introduction. It was also the same approach taken by an important collection edited by Genovese and Laura Foner at the end of the previous decade.[9]

More innovative still was the case of David Brion Davis, who, with his crucial works on "the problem of slavery," inaugurated a new way of looking at comparative slavery in the New World as an integrated and very important part of world history. Davis published three studies over the course of two decades: *The Problem of Slavery in Western Culture* (1966), *The Problem of Slavery in the Age of Revolutions* (1975), and *Slavery and Human Progress* (1984). He elaborated the first broad view, and still the most complete history, of the ideologies related to the different types of slave system that succeeded one another during the development of western civilization, until slavery's abolition. [10]

Beginning with Greco-Roman civilization, Davis gradually arrived, through the selection of particularly significant episodes, to treat the nineteenth century, showing clearly the degree of continuity in the ideologies that justified slavery from the ancient world to the modern world. According to Davis, the crucial element that provided this continuity was the notion of progress—a notion that, until the second half of the eighteenth century, was inextricably linked to the existence of slavery. Until then, slavery was considered not only a normal condition, but even the most efficient and most modern labor system. Davis was the first scholar to show how the influence of the Enlightenment and of the doctrine of natural rights, together with the spread of the classical capitalist notion of "free" labor, led intellectuals and economists to consider slavery, for the first time in its long history, as an inhuman and backward labor system.

Equally influential on the historiography of comparative slavery have been three different works that were early examples of broad types of analyses of what is now called *world history.* The first of these works was Orlando

Patterson's *Slavery and Social Death* (1982). It was specifically focused on slavery and became quickly the privileged model for world histories of slavery in general. Engaging in a broad comparison of hundreds of slave societies in different regions and at different times, Patterson aimed at arriving at a definition of the basic constitutive elements of slavery that had a universal value. He formulated the now famous expression of "social death," which indicated the special status of property that characterized slaves. This was a status that placed them outside the social structures that regulated the life of free individuals.[11]

The second work was Immanuel Wallerstein's *The Modern World-System* (1974–2011). This was essentially an economic history of the world in the modern period, in which the author argued that, starting from the sixteenth century, a capitalist world-system dominated by Europe began to exist. Within this world-system, according to Wallerstein, the stronger center in northwestern Europe was characterized by incipient industrial capitalism and free labor, while the peripheries produced raw materials for the center and were characterized by mostly unfree forms of agricultural labor, or else by sharecropping.[12]

The third study was Eric Wolf's *Europe and the People without History* (1982). It essentially dealt with the history of the forms of exploitation, paramount among which was slavery, endured by the indigenous people of different parts of the world as a consequence of European expansion. Similar to Wallerstein's case, Wolf's central theme was the rise of modern western capitalism, which, however, he placed in the nineteenth century. During this period, the transformation of the world market led to a global acceleration in the process of economic change. This process reached a particular intensity in preeminently agrarian regions characterized by the presence of large landed estates.[13]

Also very important, and very much in dialogue with the previously cited works, was the parallel growing influence of the new scholarship on the *middle passage*—the name given to the trade in slaves taken across the Atlantic from Africa to the Americas—whose initiator was Philip Curtin. With his seminal study, *The Atlantic Slave Trade* (1969), Curtin had effectively written the first comparative treatment of the *middle passage,* analyzing the role of the European colonizing powers, as well as looking at it from the African and American perspectives. Curtin's main focus was on statistics and the volume of the trade. However, his study provided a model for subsequent studies such as Herbert Klein's comparative monograph *The Middle Passage* (1978), which engaged in an increasingly deeper and more complex comparative analysis of the different types of Atlantic slave trades and their effects on Africa and the Americas.[14]

As a result of all these developments, by the early 1980s, all was in place for slavery in the Americas—the study of the totality of the experiences of slave societies in the New World—to rise to the fore within the emerging historical discipline of Atlantic history—the study of the history of the societies that created an integrated Atlantic world between the sixteenth and nineteenth centuries. Though it was going to be still a few years before Atlantic history became an actual historical field in its own right, from then on, the main ideas at the heart of that discipline clearly informed all the most important studies of slave societies in the United States, Latin America, and the Caribbean. In the process, these ideas also provided an innovative, flexible, and well-devised, framework within which both rigorous and soft comparisons of slave societies in the New World could fit effortlessly.

FROM THE 1980S TO THE PRESENT: BEYOND ATLANTIC SLAVERY

In the past 20 years, thanks to an impressive body of scholarship made of both interpretive monographs and path-breaking collections of essays, slavery in the United States and in the Americas has become progressively an increasingly essential part of Atlantic history, and one with a strong type C comparative component. Important scholarly studies have looked at the entire development of slave systems in the Americas, and thus in a very wide comparative perspective, as an essential part of world history, arguing at the same time about the crucial importance of the European background to Atlantic slavery. These studies include *The Rise and Fall of the Plantation Complex* (1990) by Philip Curtin; *The Making of New World Slavery* (1997) by Robin Blackburn; *From Slavery to Freedom* (1999) by Seymour Drescher; and, more recently, *Inhuman Bondage* (2006) by David Brion Davis; *Slavery, Emancipation, and Freedom: Comparative Perspectives* (2007) by Stanley Engerman; *Abolition* (2009) by Seymour Dresher; and *The American Crucible* (2011) by Robin Blackburn.[15]

In particular, in *The Rise and Fall of the Plantation Complex,* Curtin synthesized in an especially effective way the results gained by a generation of historians, among whom he has played a preeminent role, on the origins of slave systems in the Americas from the sixteenth century onward. Curtin focused his analysis on the rise and development of a particular socioeconomic system, which he termed the *plantation complex.* This was a system based on the use of slaves and on production in large landed properties, which was present in large part in all the New World slave societies.

According to Curtin, and also to most historians now, the origins of the plantation complex stretched back to the use of slaves in sugar plantations by Venetian and Genoese merchants who, following the Arabs before them, established their business first in Cyprus, then in Crete, and finally in southern Spain and Portugal. Later, the Spanish and Portuguese exported this particular slave system to the New World at the time of the great discoveries. In the New World, the plantation complex was gradually perfected until it eventually assumed the mature features of the major nineteenth-century slave systems in the U.S. South, Cuba, and Brazil.[16]

Similar, to a certain extent, to the themes treated by Curtin is Robin Blackburn's *The Making of New World Slavery*, a particularly important study for the interpretation of the process that led to the formation of slave systems and of the ideology of slavery in the New World. In his book, Blackburn focused specifically on the European roots of the slave societies—among which particularly important were England's North American and Caribbean colonies—that flourished in the Atlantic in the early modern era, and on the influence they had on the making of European, and particularly English, economic power and imperial ideology.

Blackburn's work has contributed in a decisive way in prompting a debate among historians on the origins of those economic, social, political, and ideological elements, which we consider indicators of *modernity*, and on their relationship with slavery. Blackburn has argued that England and western Europe have followed a particular path in the creation of modernity, or the creation of a type of civilization that we recognize as markedly different from its predecessors. Not only has slavery played a crucial role in this path, but also western Europe, and England in particular, have laid on slavery their very socioeconomic and ideological foundations, as the exploitation of the Caribbean islands to the advantage of the incipient English industrial revolution, a subject already treated by Eric Williams, demonstrated.[17]

The three most recent synthetic studies of slavery in the Americas and beyond, by David Brion Davis, Seymour Drescher, and Robin Blackburn—all published in the past five years—have built on the original intuitions of Curtin's and Blackburn's earlier works in order to provide extremely informative and well-crafted overviews of the history of slavery and its abolition. In doing this, they have kept a particular focus on the New World and its connections with Europe.

Davis' *Inhuman Bondage* aimed at illuminating "our understanding of American slavery" through the analysis of its Atlantic and New World contexts. Davis looked first at the Old World origins of New World slavery, and then he embarked on a comparative study of slavery in different regions of the Americas in order to identify the institution's constituent characteristics.[18]

Drescher's *Abolition* has covered much of the same ground as Davis' book, but has focused more specifically on the eighteenth and nineteenth centuries—the heyday of the slave systems in the Americas. He has looked, in particular, at the eventual demise of slavery through emancipation, partly as a result of the rise of abolitionist movements in both the Old and New Worlds.

Blackburn's *The American Crucible* also has a great deal in common with Davis's and Drescher's studies, but, unlike the former two, it ties much more tightly the rise and expansion of New World slavery to the spread of capitalism on a global scale. Blackburn's study is particularly notable for the novelty of its transnational approach, which takes into account the crucial connections, especially in terms of the spread of ideas and principles such as the doctrine of human rights, between the Americas and Europe, particularly at the time of the French and Haitian Revolutions and the Industrial Revolution and early working-class struggles.

Though with some crucial differences, Robert Paquette and Mark Smith's edited collection *The Oxford Handbook of Slavery in the Americas* (2010) is in a similar vein. It aimed, through the contribution of a large group of scholars, to account for both the enormous diversity and also the common features of the different slave systems that flourished in the Americas.[19]

All these studies have argued, in different ways, about the novel features of Atlantic slavery's exploitative capitalist system. More restricted in focus, but still very important, are recent type A comparative studies focusing on comparisons between North American and Latin American slavery. These include Ira Berlin's and Philip Morgan's edited collection on slave labor *Cultivation and Culture* (1993), and the two monographs *Feitores do corpo, missionarios da mente* (2004) by Rafael Marquese and *The Comparative Histories of Slavery in Brazil, Cuba, and the United States* (2007) by Laird Bergad. In different ways, these monographs compared specifically the three main nineteenth-century slave societies in the Americas, whose similar features led Dale Tomich to consider them all part of a "second slavery," distinguished from the previous colonial type.[20]

Type B comparative works, which focus either on comparisons within the English-originated slave systems or on comparisons within the non-English slave systems, include syntheses such as *Questioning Slavery* (1996) by James Walvin, on slavery in the English-speaking parts of the Americas, and *African Slavery in Latin America and the Caribbean* (2007) by Herbert Klein and Ben Vinson. The latter is the second edition of a very successful survey of slavery in the Latin American countries.[21]

Together with these studies, important recent type B and C comparative collections include *Slavery and the Rise of the Atlantic System* (1993), edited by Barbara Solow; *The Atlantic Slave Trade* (1992), edited by Joseph Inikori and Stanley Engerman; *More than Chattel* (1996), edited by David Barry Gaspar and Darlene Clark Hine; *Tropical Babylons* (2003), edited by Stuart Schwartz; and *Women and Slavery* (2007) by Gwyn Campbell, Suzanne Miers, and Joseph Miller. Also as a result of the editors' and authors' strong links with the increasingly sophisticated scholarship on the Atlantic slave trade, these collections have come to represent crucial steps in the attempts at building an overall Atlantic framework within which to fruitfully compare, by means of juxtaposition, insightful case studies originating from different slave societies in the Americas.[22]

Moreover, as Atlantic history has grown as a discipline, and the Atlantic framework has become increasingly dominant among historians of slavery in the Americas, several scholars have begun to investigate thoroughly, and in comparative perspective, the African backgrounds of the slave populations of the Atlantic system. Thus, type C comparative studies such as *The Black Atlantic* (1993) by Paul Gilroy, *Africa and Africans in the Making of the*

Atlantic World (1998) by John Thornton, *Saltwater Slavery* (2007) by Stephanie Smallwood, and *Slavery and African Ethnicities in the Americas* (2007) by Gwendolyn Midlo Hall have done much to place the representatives of different African ethnicities at the heart of those momentous transformations that, in comparable ways all across the Americas, gave origin to an Atlantic system of slavery. [23]

While Paul Gilroy has argued convincingly in favor of the existence of a shared black Atlantic culture, John Thornton has looked at the origins of that same culture, analyzing the African societies' and elites' involvement in the early Atlantic world. Then, building on Thornton's pioneering work, Smallwood has recently written about the African perspective on the Atlantic slave trade and the latter's role in erasing and reconstituting African identities, while Midlo Hall has focused specifically on the persistence of African identities among particular ethnic groups that were brought to the New World.

Since the beginning of the 1980s, a smaller group of scholars, all of them comparative historians, has argued in favor of an even wider perspective than that of Atlantic studies, pointing out the comparability between New World slavery and other types of labor, mostly unfree, spread outside the Americas. By broadening the geographical scope and employing an innovative methodology, comparison through an enlarged Atlantic perspective, including the whole of Europe and Africa, has proven a particularly fruitful terrain for the writing of explicit type C comparative studies.

These type C studies have focused particularly on slavery in the nineteenth-century American South and unfree labor in South Africa, as in George Fredrickson's *White Supremacy* (1981); or Russian and Prussian serfdom, as in the case of Peter Kolchin's *Unfree Labor* (1987) and Shearer Davis Bowman's *Masters and Lords* (1993); or sharecropping and tenancy in southern Italy, as in Enrico Dal Lago's *Agrarian Elites* (2005). Moreover, this enlarged perspective has generated a debate on the nature of slavery and serfdom in world history, which in turn has produced highly valuable comparative works, much wider in scope. Particularly important among these wider comparative studies are *Servitude in Modern Times* (2000) by Michael Bush, and *Through the Prism of Slavery* (2004) by Dale Tomich. These have argued in favor of the comparability of New World slavery with an ever-wider range of case studies of unfree labor in different regions and at different times.[24]

CURRENT TRANSNATIONAL AND COMPARATIVE TRENDS

Currently, scholarship on American history, the U.S. South, and New World slavery is in the midst of a "paradigm shift" in historiographical terms, as a result of the increasing popularity of transnational ideas and approaches to research and writing. As several scholars have pointed out, the transnational turn arose in the United States as a result of the efforts by a small but influential group of scholars—among the best known are Ian Tyrrell, David

Thelen, Thomas Bender, and Carl Guarneri—to put to rest the old idea of American exceptionalism (in short, the supposed difference of the United States from other nations due to a number of different factors).

In a 1991 issue of the *American Historical Review,* in the first of a series of forums dedicated to the "new transnational history," Tyrrell asked American historians to direct their research to the study of "international connections," focusing—in the words of Michael McGerr—on "instances of the impact of international trade, migration, reform movements on the United States," as well as on the American contribution to the shaping of a global society and environment. Sixteen years later, in the introduction to his 2007 book *Transnational Nation,* which was a transnational overview of American history, Tyrrell wrote. "What is transnational history? It is the movement of peoples, ideas, technologies, and institutions across the border." Criticized for not giving enough importance to the nation, Tyrrell, then, wrote categorically that "transnational does not mean that the nation is unimportant ... [but] the 'national' must not be assumed."[25]

Interestingly, a few years earlier, in a 1999 issue of the *Journal of American History* entirely dedicated to transnational perspectives, David Thelen, a convinced transnationalist, had argued that "a transnational method can bring the nation-state itself into sharper focus," especially if combined with historical comparison, rather than do away with the nation-state altogether. For his part, both in a seminal 2002 edited collection and in his own transnational history of the United States, called, significantly, *Nation Among Nations* (2006), Thomas Bender has done much to place national themes and issues of American history, among which paramount have been both slavery and the Civil War, within an international framework apt to a truly global age. Since it is transnational and global at the same time, Bender has also called his approach *cosmopolitan,* as the United States has been a "cosmopolitan" nation for the entirety of its history. Bender has urged scholars "to notice the evidence of transnationalisms previously overlooked or filtered out by historians."[26]

In their plea for internationalizing and globalizing American history, U.S. historians such as Tyrrell and Bender have influenced the historiographical debate on many disciplines in many other countries, including slavery studies. Now the transnational turn is a reality, firmly established among English-speaking historians all over the world.

In general, though, the transnational turn in both the United States and Europe has generated a host of criticism toward comparative history. Two main objections to the comparative historical method seem to have arisen from both English-speaking and continental European transnational historians.

The first objection is that, as Jurgen Kocka, a premier German comparative historian, has noticed, "comparison appears a bit too mechanistic, a bit too analytical in that it separates reality into different pieces in order to analyze, that is to compare, the pieces as units of comparison, whereas it would be necessary to see them as one, as one web of entanglement." The second most important objection is that, in the words of Charles Bright and Michael Geyer, "the more substantial problem with the attempt to internationalize ... history through comparative studies is that it treats societies as discrete, free-standing entities, and it tends to take the nation as a presumptive and pre-existing unit of containment at the center of the story."[27]

Presently, in both the United States and Europe, after a period of heated debates, transnational and comparative scholars seem to be coming to a consensus of a sort, recognizing that it is possible to mediate between the two methods. For example, Heinz-Gerard Haupt and Jurgen Kocka have claimed that "the study of entanglement [that is, of relationships and connections between regions and nations] ... can profit from comparative history's methods." This echoes similar recent statements expressed by Benedicte Zimmermann and Michael Werner, who are both staunch supporters of two transnational methods of historical analysis called *histoire croisée,* or "entangled history," and *transfergeschichte,* or "transfer history." Both methods focus on social and cultural connections and/or ideological transfers between nations. At the same time, in the United States, Ian Tyrrell has envisioned "the linking of comparative study of settler societies (large areas of the globe where whites came to dominate) with transnational contexts of imperial power and the expansion of global markets under capitalism." [28]

What these positional statements seem to tell us is that transnational and comparative history could be combined together in a world history perspective—one that looks at the global entangled history of past societies, and effectively compares them with one another. This is a particularly important point to bear in mind when looking at American slavery and slavery in the Americas in comparative perspective.

Would this be possible? Are transnational history and comparative history truly reconcilable within the framework of a world or global history perspective? And how would that affect comparative studies of American slavery and slavery in the Americas? To answer these questions, we must first define what world history and global history are and the challenge they pose to the comparative historical method.

Much like transnational history, the *new world history,* which distinguishes itself from previous attempts to write a total history of the world, began in the 1990s. From the beginning, it had much in common with the transnational turn, as the studies on the history of American slavery in world perspective cited in the previous pages can testify. As Charles Bright and Michael Geyer wrote in 1995, "The central challenge of a renewed world history ... is to narrate the world's pasts [plural] in an age of globality." In 2009, after 14 years during which the field had grown and defined better its scope, Marcus Gräser could write "world history is shared history." Rather than being concerned with the entire world, it considers the world as it is present in and constituted by the elements of "interconnectivity, large and small."[29]

Thus, world history nowadays is as close as it can be to a sort of transnational history on a global level. It is very much concerned with similar issues of the flow of goods, people, and ideas, and their effects on the history of different areas and civilizations, but the scale and the declared scope of world history is dauntingly larger. This is especially true of current world historical studies of slavery, in which the emphasis is specifically on the interconnections between different continents within a shared framework of cross-oceanic enforced diaspora, caused by the need for unfree labor—whether this diaspora occurred in the Atlantic, Indian, or Pacific Ocean.

Can comparative history, and specifically the comparative study of slavery, effectively meet the serious challenge posed by world/global history of integrating the global with the local in a transnational study? I believe we can begin to find hints to the answer to this question by looking at what some scholars have called *cross-national history,* as opposed to transnational history.

It all began in 1995, when, in the *Journal of American History,* renowned comparative historian George Fredrickson made a strong plea for "cross-national comparative history," by which he meant a comparative historical approach to the study of the past conducted across the world's nations in transnational fashion. In George Fredrickson's own words, "cross-national comparative history can undermine two contrary, but equally damaging presuppositions—the illusion of total regularity and that of absolute uniqueness."[30]

In Fredrickson's view, by employing cross-national comparative history, we have the potential to defeat once and for all, on a world scale, ingrained concepts of exceptionalism, which plague the historiography of every country. Moreover, according to Fredrickson, "cross-national comparative history encourages interdisciplinary perspectives" on the history of the world, since the proper way to do it is through collaboration between historians, sociologists, and political scientists. He thus alerted historians to the "theoretical attention" they should give to, and the questioning attitude they should keep in respect to, "the meanings of such analytical categories as slavery, race, class, gender, urbanization, government, nationalism, and social movements."[31]

Expanding on Fredrickson's original intuition, in the introduction to their 2004 seminal edited collection *Comparison and History: Europe in Cross-National Perspective,* Deborah Cohen and Maura O'Connor have argued that "cross-national histories follow topics beyond national boundaries ... seek to understand reciprocal influences." Thus, "understood broadly, cross-national history includes the history of colonialism and imperialism," and clearly also the history of slavery. More than this, it also includes, though it's not limited to, what we have called so far *transnational history.*[32]

Cautioning the reader who might be confused about the two terms *cross-national* and *transnational,* Cohen and O'Connor stated that, "if transnational analysis presupposed a skeptical stance towards the nation as the chief organizing agent of history, we intend 'cross-national' by contrast as a more neutral term to describe the scope of a historian's investigation." In a perceptive essay in the same collection, Michael Miller argued that "ineluctably, cross-national history takes on a comparative dimension" and that "at their best, cross-national and comparative histories are complementary, rather than competing methods of writing multinational history, and it is difficult to write one without incorporating the other." According to Miller, "whereas comparative history's strength is its ability to explain varying patterns, cross-national history's advantage is its ability to reach beyond what could be accomplished through comparative history alone."[33]

Thus, taking on board Cohen's, O'Connor's, and Miller's perceptive comments, we may argue that cross-national history could prove able to combine together the transnational and the comparative approach—a combination much advocated by scholars working with both methods—into a cross-national comparative history on a world scale of the type envisioned by Fredrickson. Put simply, the cross-national comparative history approach would transcend the nations in attempting to understand the connections between them on a world scale, but at the same time, also respect the nation as a possible unit of analysis when it came to comparison.

If we took this line of reasoning a step further and we tried to integrate fully this type of cross-national comparative history within the larger and, in some respects, daunting framework of world/global history, we would need to adapt it somehow. In that case, we could not simply restrict ourselves to nations (also given that they are, relatively speaking, latecomers on the world history scene) and to synchronic studies of broadly contemporary phenomena. Instead, we should envision a broader approach combining the comparative method with the study of both synchronic and diachronic connections of phenomena such as slavery—thus, also those connections that related to different historical periods—into a type of cross-continental comparative history. This would not necessarily mean that continents would supply the scale of study, even though they obviously could do so, but rather that we would look for both comparisons and connections across continents, both diachronically and synchronically, on a world scale. An example is the study of the world history of slavery in comparative perspective within a framework that goes beyond the American continent, and encompasses both the Old World and the New World in both synchronic and diachronic fashion.[34]

As it happens, a cross-continental comparative and diachronic approach is the type of approach that C. A. Bayly, being extremely critical of the applicability of the word *transnational* to the world before 1914 (when, in his view, modern nation-states had not appeared yet), has favored in the most celebrated single-volume recent study of the long and global nineteenth century: *The Birth of the Modern World, 1780–1914* (2004), significantly subtitled *Global Connections and Comparisons.*

In his book, Bayly has sought to investigate the birth of modernity on a global scale. Rejecting once and for all both the old Eurocentric view and the more recent Sinocentric perspective, Bayly has considered civilizations present in different continents in the nineteenth century as equally important constituents of modernity. In order to do this, Bayly looked transnationally at the "rise of global *uniformities* in the state, religion, political ideologies, and economic life as they developed throughout the nineteenth century." He has claimed that, while these moves toward connections created "hybrid polities, mixed ideologies, and complex forms of economic activity ... [they] could also heighten the sense of *difference,* and even antagonism, between people in different societies, and especially between their elites." [35]

In his work, Bayly has taken an approach that has effectively integrated the transnational view with specific comparisons across different regions, states, and nations by focusing on broad issues, including slavery, serfdom, and free and unfree systems of agrarian labor. The resonance of these issues in the nineteenth-century world, in fact, was global, or, at the very least, cross-continental.

At the end of this long reflection on current historiographical trends related to the transnational turn and world history, and their impact on comparative history and the history of comparative slavery, we can say that it is possible, and also commendable, using C. A. Bayly's *The Birth of the Modern World* as a model, to integrate transnational and comparative approaches into a single coherent approach to the history of American slavery in comparative perspective.

This integrated approach is particularly well suited to connect the global to the local, and therefore it answers the question of whether transnational history and comparative history are reconcilable within the framework of a world or global history perspective. It does so by acknowledging the world historical significance of movements of goods and people across regions and continents, and at the same time recognizing the historical peculiarities of these very regions through a comparative perspective geared toward understanding the similarities and differences between them.

In particular, on one hand, the transnational and world history view is particularly useful in showing links and connections between slavery in the American South and other slave societies in the Americas, and also in Africa and in a wider Atlantic world that saw continuous exchanges with western Europe. This was especially the case in the seventeenth and eighteenth centuries, and in the revolutionary period, because of the nature of colonial empires. However, as an increasingly larger number of scholars currently acknowledge, in the nineteenth century, links between nations and regions of the Atlantic world not only continued to exist, but even intensified.[36]

On the other hand, the comparative approach is particularly useful in the combined analysis of the American South and other societies, such as regions of eastern and southern Europe, which had few contacts with the American slave system. However, these are still worth investigating in order to shed light on similarities and differences between systems of labor—free and unfree—radical movements, and processes of emancipation and national consolidation.

In short, while the transnational approach is more useful in investigating in comparative perspective American slavery in its New World and Atlantic dimension, the comparative approach is more useful in investigating American slavery within its wider Euro-American framework. Even though mostly concerned with the actual comparison between American slavery and other slave societies and systems of unfree labor, the study presented in this book seeks, to a certain extent, to combine the comparative approach and the transnational approaches by using them, whenever possible, in both sections of the book—Part A: American Slavery in the Atlantic World and Part B: American Slavery in the Euro-American World.

From the first studies in the 1940s to the latest scholarship, comparative studies of American slavery and slavery in the Americas have faced increasingly difficult challenges. On one hand, these challenges were related to the continuously deeper understanding of the workings of New World slave societies by subsequent generations of slavery scholars. On the other hand, the increasing awareness of the crucial importance of the wider context—the Americas, the Atlantic world, and the Euro-American milieu—was pointed out by an ever larger and more complex group of works. Yet, the comparative approach to American slavery maintains its validity as a unique method focused on the discovery of the similarities and differences between different slave societies and also between societies characterized by a variety of types of unfree and free labor systems. If combined with the transnational view and method, with its emphasis on particularly important links and connections between different regions and at different times, the comparative approach to American slavery has the potential to shed light and yield further insights on the history of the U.S. "peculiar institution" that would not be possible to discover otherwise.

Part A

AMERICAN SLAVERY IN THE ATLANTIC WORLD

CHAPTER ONE

AMERICAN SLAVERY AT THE PERIPHERY OF THE ATLANTIC SYSTEM

Comparison has proven crucial in our understanding of the origins of New World slavery. By investigating similarities and differences across time and space, comparative historians have established the extent of the continuity and change that slavery in the Americas has represented in world history. Diachronic comparisons with ancient and medieval Europe have shown that Atlantic slavery in the New World was the first continental scale and utterly pervasive slave system, both economically and socially, since the Roman Empire. And yet it differed from the latter, as it differed from all previous slave systems, in being simultaneously capitalist and centered on racial exploitation.[1]

According to Michael Bush, unlike ancient or medieval slavery, "the slave system brought by the Europeans [to the Americas] was hereditary, racist, and designed to generate wealth, notably through the mining of precious metals and the cultivation of cash crops." Through diachronic comparisons, it has been possible to trace the origins of both American racism against Africans and also the particular systems of labor implemented in New World plantations to their late medieval European antecedents. Equally, diachronic comparisons have also enlightened us on the long story of the different crops—sugar, rice, and tobacco—that came from the Mediterranean and different areas of the Atlantic, notably Africa and the Americas themselves, to make the early fortunes of plantation slavery in the New World.[2]

Yet, it is synchronic comparison that has told us more about the origins of New World slavery between the sixteenth and seventeenth centuries. Comparisons between different slave societies in the Americas, and between American slavery and eastern Europe's so-called *second serfdom* have proven incontrovertibly that the single most important factor that led to the making of a massive and highly racist Atlantic slave system in the Americas, as well as to the rebirth of a massive serf system in eastern Europe, was a shortage of labor and an availability of land at a time of large-scale economic expansion. In particular, in the Americas, mostly Mexico and Peru had large native and agriculturally skilled populations, while areas such as Brazil, the Caribbean, and mainland North America had sparse native populations that were not nearly as agriculturally advanced. In these areas, crop cultivation through plantation slavery became quickly endemic. And, as David Eltis has shown, when the employment of European white servitude became too expensive in those New World areas, the only option left to the planter elites was the transatlantic slave trade of Africans.[3]

The Atlantic perspective of different comparative studies has been crucial in understanding how the different planter elites of the New World coped, in both similar and different ways, on one hand with the constantly changing conditions of the world economy, and on the other hand with their equally constantly shifting status of suppliers of highly requested raw materials at the periphery of global European empires. The Atlantic perspective also has been critical in clarifying how, enslaved and taken by European slave traders from a variety of societies that were at different stages of agricultural and technological advancement, the West and Central Africans of different ethnicities who survived the lethal *middle passages*—the forced transatlantic migrations—of the various Atlantic slave trades managed by European powers brought with them features of their original African cultures (see Map 1). Some of these cultures had been long in contact with Europeans, and this had a profound influence on the developing of the Atlantic slave system. Among those features that African slaves brought with them to the different plantation areas of the Americas was their invaluable knowledge of plant cultivation. This gave a crucial contribution to the making of New World slave economies and societies, most notably in the case of rice.[4]

ATLANTIC SLAVE TRADES AND MIDDLE PASSAGES

Historians have estimated that between the mid-fifteenth century and the end of the nineteenth century, Europeans moved forcibly from Africa to the Americas a staggering number of more than 12 million slaves. The majority of the slaves that were brought to work in the New World plantations came from the West Coast of Africa, the region around the Gulf of Guinea, present-day Nigeria, the Gold Coast and Ivory Coast, and Senegambia. These areas were characterized by tropical vegetation and were covered with forests that alternated with savannahs.

Traditionally, African societies were based on the centrality of the institution of the family and on the existence of small villages characterized by a strong sense of belonging to a community. The activities of the members of families and villages were related to the agropastoral economy that characterized the entire area of the Gulf of Guinea.[5]

In several regions, the villages were parts of sociopolitical institutions of different sizes. Some of these were kingdoms and empires that fought against each other for power over certain areas. Among the most important were the Mali and Songhay empires (which between the thirteenth and seventeenth centuries succeeded one another in the exercise of hegemony over vast areas in West Africa); the Kingdom of Benin, in present-day Nigeria; and the Kingdom of the Kongo.[6]

As a result of the wars that involved periodically both great kingdoms and small villages, prisoners were often captured and made slaves. However, this type of slavery, which characterized all traditional African societies, was closer to domestic serfdom.[7]

The prisoner case was different from the case of those slaves whom, from as early as the eighth century AD, Arab merchants sold to African rulers. In fact, during the medieval period, the Arabs utilized a large enslaved workforce, made of both Africans and Europeans, for agricultural labor in their plantations throughout the Mediterranean.[8]

By the twelfth century, the Normans in Sicily and the Europeans ruling the Crusader states in the Levant had managed to learn from the Arabs the technique of cultivation of the sugarcane, and they used it with great profit in large landed estates, exploiting a workforce made of both slaves and serfs. In time, sugar cultivation and slavery became inextricably linked to one another, and the Venetians practiced both, first in Crete and then in Cyprus, while both Venetians and Genoese merchants were in control of a vast Mediterranean slave trade. By the first half of the fifteenth century, supported by the Spanish and Portuguese monarchs, and thanks to the new and flourishing trade with the great kingdoms of West Africa, Genoese merchants managed large sugar plantations in the southern regions of the Iberian Peninsula, whose workforce was mostly made of African slaves.[9]

After establishing sugar plantations on several Atlantic islands in their possession—notably Madeira, which by the 1480s was the leading world producer of sugar—at the beginning of the sixteenth century the Portuguese brought the practices of sugar cultivation and plantation slavery, with preeminently African labor, to their new colony of Brazil. From there, these practices extended to the Spanish colonies and eventually to all the Caribbean islands. In the course of the seventeenth century—as a result of the activities of Dutch, French, and English merchants and pirates—the center of sugar cultivation and the focal point of the now-established Atlantic slave trade between Africa and the Americas moved to the Caribbean, within the context of a European economic system with global dimensions. Therefore, mostly as a result of the rapidly growing demand for sugar from Europe, different European powers engaged in their own trade in order to bring enslaved Africans to work in their plantations located in various areas of the Americas.[10]

Map 1 Atlantic Slave Trades and the African Origins of Slaves in the Americas

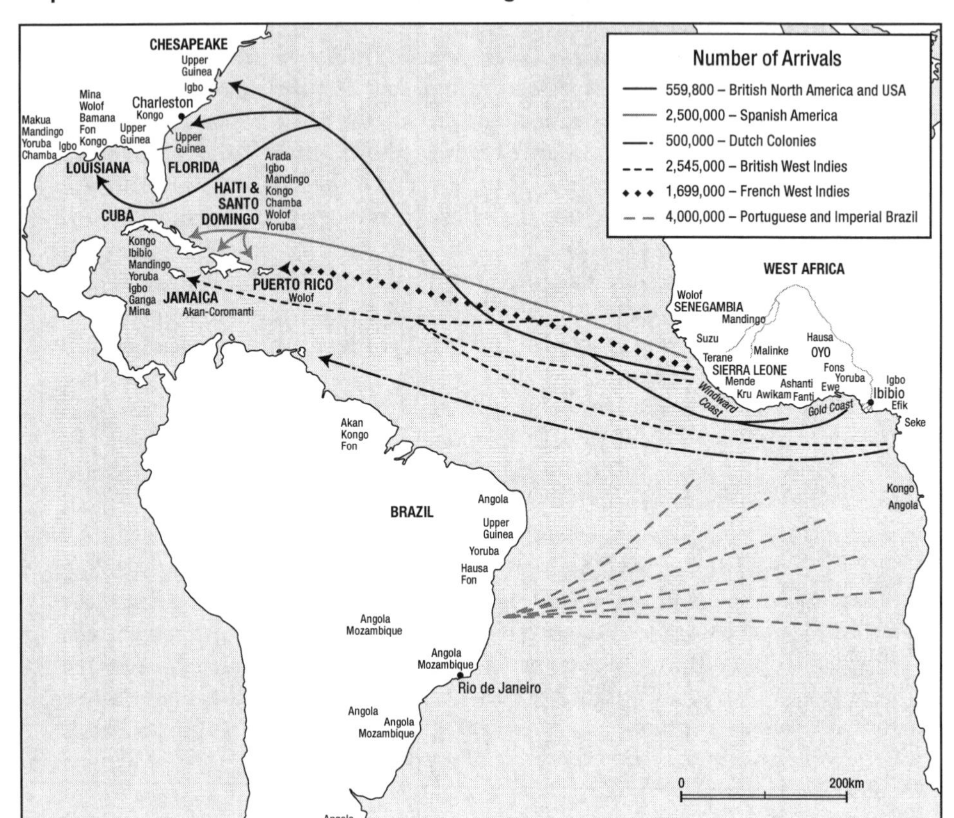

Therefore, it is correct to say that the Atlantic slave trade saw different middle passages managed by different European merchants. Even though from the point of view of Africans and African Americans, all the slave trades had similar features and produced similar results, their history and historical significance differed, as did the volume and importance of their trade. Ultimately, a comparison can help elucidate the reasons why by the beginning of the eighteenth century the British slave trade—only a small fraction of which served the colonies of the American South—became the most important one, and remained so for just a century, until its abolition in 1807.[11]

Iberian Slave Trades

Together with Spain, the longest profiting European power in the Atlantic slave trade was Portugal. This was thanks to the earliest efforts at Portuguese exploration, which earned them their first African colonies, both in the Atlantic islands and on the mainland. By 1482, when they founded their most important trading post of São Jorge da Mina (Elmina) in Ghana, the Portuguese were in a position of monopoly in regard to the African trade of not just slaves, but also of gold and ivory. In the sixteenth century, the Portuguese reinforced this position with the settlement of the African colony of Angola and with the creation of the American colony of Brazil.[12]

Relying heavily on the African indigenous tradition of slavery and internal slave trade, the Portuguese Atlantic slave trade functioned with a model according to which "the king of Portugal granted ... monopoly trade privileges to Portuguese merchants and nobles regarding a certain portion of the African coast for a certain period of time." The grant holders would then usually sell some of their shares, effectively subcontracting the business to merchants who acquired and transported African slaves. Through this system, the Portuguese created one of the first integrated economic zones in the Atlantic world. The zone was held together by the logics and practices of the slave trade between Africa and the Americas, and initiated many of the standard horrific practices of slave transport on their slave-trading ships in the earliest of several European versions of the middle passage.[13]

As a result of competition from other European powers, after the early sixteenth century, the Portuguese lost their monopoly. However, the volume of their Atlantic slave trade continued to increase, reaching its peak in the eighteenth century, with most of the slaves brought from Africa to the highly profitable sugar-producing colony of Brazil. There, the total number of slave imports rose from 610,000 in the period of 1451 to 1700 to 1,498,000 in the period of 1701 to 1800.[14]

From the sixteenth century until the seventeenth century, Spain was the leading European power involved in the Atlantic slave trade. Spain was also the first European power to practice the slave trade on an Atlantic scale, as a result of its acquisition of a large territorial empire in both the Caribbean islands and mainland Central America and South America in the early decades of the sixteenth century.[15]

The consequent exploitation of land and resources in the new colonies went hand in hand with a growing demand for labor in the ubiquitous agricultural enterprises known as *encomiendas*. These were land and tribute grants given by the Spanish crown to entrepreneurs deemed responsible for indigenous workers who supplied the required labor. The demand for labor grew especially after the introduction of the first sugar mills in 1509.[16]

Starting from 1518, Spain began its organized transport of slaves through an Atlantic trade and its own version of the middle passage. Most of the slaves first went to the Caribbean islands, but then also to the mainland American colonies from Mexico to Chile, arriving in gold mines or on sugar plantations.[17]

Similar to the Portuguese system, the Spanish system was also based on the government, specifically Spain's *Consejo de las Indias* (Council of the Indies) in this case. The council granted several licenses or contracts to individual traders for the supply of slaves to the Spanish-American colonies. Starting from 1595, the Spanish crown granted a single "monopoly contract," called *asiento,* to those individuals, joint-stock companies, or states, which would effectively import a certain number of slaves to the Spanish-American colonies.[18]

Despite the Spanish intention to create a monopoly, as David Brion Davis has pointed out, "not only were *asientos* sold and subcontracted, [but also] privateers and interlopers from various European countries continued to break any meaningful monopoly" that the Spanish might have. The Spanish were forced to grant the *asiento*—which the Portuguese held at different times until 1676—to increasingly stronger European powers and former interlopers, such as France in 1702 and then Britain in 1713. As a result, the Spanish eventually dismantled the *asiento* system, opening the trade to all neutral nations. Altogether, in the period between 1451 and 1800, Spanish imports of slaves to Spanish America reached a number close to 1 million individuals.[19]

The Dutch and the French

The first of the state-chartered companies to attempt to break the Iberian monopoly on the Atlantic slave trade was the Dutch West Indian Company (WIC), which was founded in 1621. The Dutch WIC soon posed a serious threat, particularly to the Portuguese Atlantic economy. It managed to control a number of forts on the African Gold Coast and also to seize the island of Curaçao, which was destined to become the main Dutch slave-distribution center. The Dutch also managed to seize a large portion of the Portuguese colonies of Brazil and Angola, where about 25,500 slaves were shipped in the first 20 years of the period 1636 to 1674, out of a total of 90,000 slaves shipped overall. After the loss of Brazil in 1654, and then the 1674 bankruptcy of the first Dutch WIC, the new Dutch WIC reorganized itself and the Atlantic slave trade. But in the eighteenth century, it no longer played an equally important role as earlier on; eventually, in 1791, it was abolished.[20]

Compared to the Dutch, the French had a much more incisive influence on the politics and economics of the Atlantic slave trade. Chartered with specific rights over trade with Africa and the Caribbean, the French started with the *Compagnie des Indes Occidentales* (CIO), which lasted only from 1664 to 1674, and then the two later main French slave-trading companies named the *Compagnie du Senegal* and the *Compagnie du Guinea,* both founded in the 1670s. These were monopoly companies financed by the French state and defending its interests. The *Compagnie du Guinea* received the Spanish *asiento* in the years 1702 through 1713. Shortly afterward, however, the French-Atlantic slave trade opened to all private traders.[21]

Even though in the late seventeenth century most slaves taken by the French and brought through the middle passage ended up in the French-Caribbean colonies of Martinique and Guadeloupe, in the eighteenth century and until the occurrence of the Haitian Revolution, the majority of the slaves were taken to the island of St. Domingue, whose extremely profitable economy was equally based on sugar plantations, but on a much larger scale. Altogether, during the period of 1701 to 1800, the French-Atlantic slave trade brought almost 1.5 million slaves to the French colonies in the New World, and the overwhelming majority of that number landed in St. Domingue.[22]

The British Slave Trade to the Caribbean and North America

Although the earliest evidence of an English slave trade with Africa goes as far back as 1618, it was really in the second half of the seventeenth century that the English government made serious efforts to control the Atlantic slave trade. This was at a time when England owned a number of Caribbean islands—the most important of which were Barbados and Jamaica—and colonies in the entire Atlantic coast of North America.

To control the slave trade, the Stuarts supported the foundation of the Royal African Company (RAC) in 1672 and granted it the monopoly of trade with Africa. Even though in 1689 the RAC lost its monopoly as a consequence of the Glorious Revolution, it kept its formidable headquarters at Cape Coast Castle on the Gold Coast. From there, as a result of the prominence of the Royal Navy following the British victory in the third Anglo-Dutch War (1672–1674), it was now in a position to command the Atlantic slave trade. The RAC eventually received the *asiento* from the Spanish crown in 1713. Together with the Gold Coast, mainly the West African regions of the Bight of Biafra and Sierra Leone supplied the largest number of slaves to the British ships.[23]

Operating in triangular pattern, from British ports such as Liverpool and Bristol to the West Indies and North America and then back to Britain, during its heyday between 1701 and 1800, the British Atlantic slave trade forcibly carried on the middle passage an estimated 2,545,292 slaves destined to plantations in the Americas, according to Trevor Burnard. The booming sugar-producing economy of the British West Indies was largely responsible for the rising demand in the number of slaves, mainly as a result of the high mortality rate due to the brutal working conditions and the tropical diseases. Consequently, in the seventeenth century, Barbados received more slaves than any British colony; however, after 1720, Jamaica took its place.[24]

Notice that only a small fraction of the number of slaves taken on British ships was destined to the North American colonies. This was mainly because, in comparison with the West Indian sugar islands, these colonies were in a peripheral position *vis-à-vis* the British Empire's economic interests in the Atlantic world. Indeed, according to David Eltis, the main regions of British North America with a flourishing plantation economy—the Chesapeake (Virginia and Maryland), the Carolinas (both North and South Carolina), and Georgia—received an estimated number of no more than 140,000 slaves from Africa in the period of 1619 to 1750. Altogether, in the entire period of 1650 to 1808, colonial North America and then the United States received a total of 559,800 slaves.[25]

Comparing the different middle passages shows differences between the modes, significance, and duration of the Atlantic slave trades managed by the various European powers. However, from the point of view of African slaves, there was little distinction between the slave trades in terms of the terrible experiences they faced.

Whatever their final destination, the transport from Africa to the Americas was for all the slaves a horrific ordeal. Crammed in the belly of the ships so as to occupy all the available space, chained to one another in 0.5-meter-high decks, slaves were left in the dirt with little to eat for a period that often reached two months. Inevitably, the diseases, the exhaustion, the brutal discipline, and the suicides and rebellions provoked a very high mortality rate. Historians have estimated that 16 percent of all the slaves brought from Africa to the Americas died before being able to reach the New World.[26]

For those who survived, the arrival was just the beginning of an even more dreadful ordeal. In every major Atlantic harbor in the Americas—and in North America, particularly in the harbors of Annapolis, Maryland, and Charleston, South Carolina—slaves were taken off the ships as soon as they arrived. Then they were cleaned up and sent to large auctions. From there, slaveholders acquired them as additions to the enslaved workforce toiling on their plantations and farms.[27]

ONE CONTINENT, MANY SOUTHS

In approaching the history of the earliest New World *slave societies*—societies whose economy relied almost exclusively on slavery, as opposed to *societies with slaves,* in which slavery was present, but not central—from a comparative perspective, it is useful to take into account the concept of "many souths" elaborated by Peter Kolchin in *A Sphinx on the American Land* (2003).[28]

In his study, Kolchin referred to the "many souths" in order to point out the great internal differences and variations in terms of geography, economy, and society that existed within the slave system of the antebellum U.S. South. If we transfer this concept to a continental scale, thereby encompassing the whole of the New World, we can affirm that, within the hemispheric slave system of the Americas, there were many variations between regions that eventually saw the development of full-scale slave societies. In this chapter, we focus briefly on the development of four particular regions that underwent this process: Brazil, the British Caribbean, and Virginia and South Carolina in the American South. These four regions are perfect examples of the earliest types of what Jason Moore has termed "commodity frontiers," referring to the process of the expansion of European capitalism on a world scale and the consequent exploitation, on the peripheries of European expansion, of the environment in relation to the production of particular types of raw material, including specific crops.[29]

Slavery and Sugar Production in Early Colonial Brazil

The earliest New World colony with a plantation-based sugar production that commanded the world market was Portuguese Brazil. Sugarcane was cultivated in the northeastern provinces of Pernambuco and Bahia from the early sixteenth century. Starting in the 1540s, sugar mills called *engenhos* were built by Portuguese colonists with technology imported from the Atlantic islands of Madeira and Canary, and with funding provided by Portuguese noblemen and merchants. By the 1580s, Pernambuco had 60 sugar mills, and Bahia had 40. Combined, they accounted for three-quarters of the entire sugar production in the New World. The sugar-mill owners (*senhores de engenho*) were effectively planters with a large workforce mostly made up of slaves, who provided the field labor for the land leased to dependent cane farmers (*lavradores de cana*).[30]

Before 1570, the Portuguese colonists in Brazil relied mainly on slaves taken from the neighboring indigenous tribes. However, as a result of the Native American decimation consequent to the diseases brought from Europe, especially smallpox, beginning from the 1570s, the Portuguese turned increasingly to imports of African slaves though the Atlantic slave trade. By 1620, the Portuguese relied exclusively on African slaves. According to Stuart Schwartz, "on the average, *engenhos* in Bahia and Pernambuco had 60–70 slaves as part of the workforce, but also drew on the labor of the dependent cane farmers so that the total effective number of workers per mill was about 100–120." The exhausting pace of work, whose main activities—usually supervised by an overseer—related to cutting the sugarcane in the daytime and operating the sugar mill in the afternoon and at night, together with the abundance of diseases, resulted in a high mortality rate. This led to the continuous need to rely on new arrivals of slaves from different parts of Africa.[31]

Beginning from 1550, sugar production rose steadily in Brazil, eventually reaching its peak in the first half of the seventeenth century. By 1600, the number of sugar mills had already grown to 200. Stimulated by Europe's ever-rising demand, Brazil's sugar output in the mid 1620s was 14,000 tons per annum, with a consequent enormous increase in slave imports from Africa. The changing geopolitical situation, however, had a considerable negative effect on this general economic trend.

First, Portugal became effectively a province of Spain with the unification of the two crowns under Philip II in 1580 and until 1640. More important, after repeated attempts during the 1620s, in 1630, the Dutch WIC succeeded in capturing Pernambuco. The Dutch kept Pernambuco for more than 20 years, until 1654, when the Portuguese successfully ousted them from Brazil.

In the same years, the Dutch captured the main Portuguese slave-trading fortresses and part of Angola.[32]

The consequences of these events were multifold. On one hand, with Pernambuco gone, the main Brazilian province for sugar production for the period of 1630 to 1654 became Bahia. However, with little hope of receiving slaves from Africa, due to the Dutch control of routes and supply, planters needed to revert to the enslavement of the indigenous people of Brazil for those 20 years. The most momentous consequence, however, was the fact that, not only did the Dutch learn the techniques of sugar planting in Pernambuco, but they also brought that technology, the equipment, and the slaves to the French and British Caribbean islands. They did this first after the post-1645 drop in Pernabuco's sugar production, and then, on a larger scale, after the loss of their Brazilian foothold in 1654.[33]

Large numbers of Dutch planters left Brazil and landed with their slaves in French Martinique and Guadeloupe, and in the English colony of Barbados. Ultimately, they gave a decisive contribution to the rise of the British Caribbean as a leading producer of sugar, able to replace Brazil by the central decades of the seventeenth century.

The Seventeenth-Century British Caribbean

Starting from the late 1640s and 1650s, during the period of Dutch possession of Pernambuco and then of the Dutch ultimate defeat, the British Caribbean islands, and specifically Barbados—helped by the migration of Dutch sugar planters and the technology they brought with them—rose in importance in the world economy. By the 1670s, they became the center of world production of sugar. They maintained this position until the first two decades of the eighteenth century. Stimulated by the rising demand supported by the European aristocracy's mania for sweeteners, sugar production found an equally ideal environment and climate in Barbados as it had found in Brazil. Thus, in a few decades, sugar production in Barbados reached extremely high levels—so high that historians have talked about a "sugar revolution." This caused a continuous increase in the number of slaves transported from Africa to work in the Caribbean plantations. For the slaves, plantations in Barbados represented a true hell, since they were forced to work in inhuman conditions and unhealthy climates. The mortality rate was extremely high.[34]

Together with the high mortality rate, the continuous demand for sugar created a vicious circle by which the demand for more workforce never dwindled, but instead led to the rising importance of the slave trade from Africa. In Barbados, sugar production occurred on large plantations, whose average workforce was about 200 slaves. A relatively small planter class owned the majority of the available land, where the majority of the slaves worked. A small percentage of colonists was engaged in commercial activities, which were nonetheless connected to the plantation economy.[35]

Recently, John McCusker and Russell Menard have questioned the viability of the concept of a sugar revolution in explaining the changes that occurred in Barbados in the mid-seventeenth century. They claim that it was a slow process—lasting about 40 years and characterized, as in other plantation areas of the New World, by the planters' experimentation with crops as diverse as cotton, tobacco, and indigo—that ultimately led to the predominance of sugar agriculture. At the same time, "the integrated plantation came slowly to replace the dispersed method of organizing sugar production." At the end of this slow process, altogether there were only a few hundred planters in Barbados. The slaves who worked in the sugar plantations, instead, made up from 60 to 90 percent of the entire population of the island.[36]

As a result of the continuous arrivals of slaves from Africa and of the replacements due to the high mortality rate, during the period between the mid-seventeenth century and the mid-eighteenth century, the Barbadian workforce was formed only in minimal part by *seasoned slaves*—that is, slaves who were already adapted to the conditions of the New World. It was mostly composed of slaves coming directly from Africa and not yet seasoned. In general, in Barbados, slaves worked divided in gangs in sugar plantations that were much larger than the average size of other plantations, especially North American ones. For this reason, it was much more common among Barbadian planters than among North American planters to leave slave supervision and plantation management in the hands of one or more overseers, even though this decision might often lead to problems due to the overseer's incompetence or cruelty toward the slaves.[37]

Slavery and Tobacco in Early Colonial Virginia

Compared to either Brazil or Barbados, Virginia and South Carolina—the first English colonies in North America characterized by plantation agriculture—played a minor role in the world economy. Still, similar to sugar, by the mid-seventeenth century, Virginian tobacco and South Carolinian rice commanded their own particular niches within the world market. Recent research also suggests that plenty of links existed between Barbados and Virginia and South Carolina as part of an "extended Caribbean" area of cash-crop cultivation associated with plantation slavery.[38]

The early years of the first permanent English colony in Virginia—Jamestown, on the Chesapeake Bay—were plagued by famine, disease, and continuous conflict with Native Americans. Then, from 1616, John Rolfe began growing tobacco. In a few years, a tobacco-chewing fashion took Europe by storm, creating a rapidly increasing demand for the crop and leading to the rise of Virginia as the largest tobacco-producing region in the world. Later, between 1650 and 1670, the crisis of the English aristocracy led a number of gentry families to migrate to Virginia, where they engaged in tobacco cultivation on a grand scale in large plantations. Yet, even though in Virginia every landowner was called *planter,* the major planters were only the ones who had large enough farms to be called *plantations.* On these plantations, they kept a workforce made of both African slaves and white indentured servants—men and women who had fled England and paid the price of the transatlantic voyage by working on someone's plantation for an average of 10 to 20 years.[39]

Even though in seventeenth-century Virginia land was abundant and cheap, it was worth little without slaves and servants to work it. Already at this time, Virginian society was like a pyramid, with slaveholders at the top and the few planters at the pinnacle. At least until the 1670s, land was readily available also because there were few free individuals who could own a piece of it. The majority were indentured servants, and the harsh conditions led to the death of many of them before the end of their term, leaving only a few to enjoy freedom.[40]

In 1676, Nathaniel Bacon led a group of rebels, mostly made up of indentured servants but also including some African slaves and a few small farmers, against the Virginian planter elite. They managed to set Jamestown on fire, but the governor Lord Berkeley subsequently crashed the rebellion, hanging 33 rebels. After this episode, also as a result of the increasingly lower number of indentured servants coming from England, the tobacco planters of Virginia were much more careful in seeking support from the lower class of whites. At the same time, due to a host of reasons, the number of white indentured servants declined rapidly, while, already with the foundation of the RAC in 1672, England had begun organizing a massive slave trade from Africa to its American colonies. This led to a huge rise in the number of Africans arriving in Virginia and to their eventual unchallenged prominence as the preferential workforce on tobacco plantations.[41]

Slavery and Rice in Early Colonial South Carolina

Compared with Virginia, the origins of South Carolina—the other region of early British colonization in the American South—include particular features that show much better the Atlantic connections of American slavery with the type of slavery practiced in the other British colonies. It particularly demonstrates the contacts and exchanges between both the master classes and the enslaved workforces of the different areas of the Atlantic world.

South Carolina's early colonization began in 1660, when Charles II gave a large land concession in the southern territory of the Carolinas to a few of his aristocratic supporters, soon to be called *Lords Proprietors.* Soon after, planters with their slaves migrated from the British Caribbean colony of Barbados, where there was a major problem of overpopulation. Moving from one side to the other in the British Atlantic world, several Barbadian planters looked for a place to replicate the planter economy and society, and finally found it in South Carolina.

Along with the Barbadians, a group of settlers from another area of the Atlantic world—the Huguenot refugees escaping from Louis XIV's France—arrived during the same decades. By the late seventeenth century, both groups had become part of the relatively small but very powerful planter elite of the British colony.[42]

The fortunes of South Carolinian planters were tied to the discovery that the subtropical climate of the colony, though not apt for the cultivation of sugar, was perfect for the cultivation of rice. The process actually took a couple of decades. The planters made several attempts until, entirely thanks to the contribution of the African slaves they brought with them from Barbados—a particularly telling tale of the importance of the Atlantic connections for the origins of American slavery—they discovered the perfect suitability of South Carolina Lowcountry's coastal swamps for rice. Along with that, they also imported the techniques of cultivation associated with rice, particularly the creation of water pools for growing rice plants.[43]

Since, in order to grow rice, it was necessary to have the initial capital to make the required infrastructure and also to own the necessary number of enslaved African workers to put on relatively large plantations, the end result was that those who began the business of rice cultivation tended to be already wealthy individuals, such as the migrated Barbadian planters and the Huguenot refugees. Throughout the seventeenth century, the constantly rising market demand ensured that rice cultivation was a very profitable business. By 1690, South Carolina was the largest rice-producing region in the world. The city of Charleston, founded in 1670, had rapidly become not only the favorite residence of rice planters and a very important market for the export of rice, but also one of the most important ports for the import of African slaves in the entire Atlantic world.[44]

Comparison highlights the fact that, in each of the four regions of Brazil, the British Caribbean, Virginia, and South Carolina—at once many souths and commodity frontiers—at different times and for different reasons the focus was on the cultivation of a particular crop that was able to command a particular niche within the world market—whether it was sugar (the most sought after by Europeans), tobacco, or rice. This led to the transformation of the initial European settlement, in a relatively short time, from a society with slaves to a full-fledged slave society. The powerful planter elite sat at the top of the economic and social hierarchy, supported by a large population of enslaved Africans supplied by the ever-increasing volume of the Atlantic slave trade.[45]

Comparison shows how, despite differences due to a number of factors, during the course of the seventeenth century, this process followed a substantially similar pattern both in plantation areas such as Brazil and the British Caribbean, whose sugar production was at the heart of Atlantic economic exchanges, and in essentially peripheral areas, such as Virginia and South Carolina. Here, production and trade of tobacco and rice was flourishing, but was peripheral in comparison with the massive Atlantic sugar network.[46]

BETWEEN THE AFRICAS AND THE AMERICAS

Slaves came to work on the first plantations in the Americas from different regions of Africa, characterized by different cultures and languages. Depending on which European power—whether the Portuguese, Dutch, French, or English—managed the slave trade in the region and at what time, slaves would be supplied by the following areas, mostly located in West Africa: Greater Senegambia/Upper Guinea, the Ivory Coast, the Gold Coast, the Bight of Benin, and the Bight of Biafra. Slaves also came from West Central Africa, Mozambique, and even Madagascar.

As a consequence of the different European powers' reliance on particular African regions for their Atlantic slave trade supply over a long period of time, often slaves from a particular area of Africa would be taken for a number of years to the same plantation area of the New World. As a result, especially in the seventeenth and eighteenth centuries (when all the major European powers practiced the Atlantic slave trade on a large scale), the chances for what Gwendolyn Midlo Hall has termed a "clustering of ethnicities" were very high throughout the early slave societies of the Americas. The consequences of this process in terms of transplantation of African cultures and of their continuing influence in the New World were incalculable, partly in contrast with what Sidney Mintz and Richard Price had argued in an influential 1976 study.[47]

In practice, African cultural identity remained strong, even though subject to a process of *creolization,* or adaptation to the Americas and to the contact with different types of white culture. This was particularly true among those ethnicities that arrived in clusters and in different waves in the same parts of the New World, as studies of African-led slaves rebellions in the Americas in the seventeenth and eighteenth centuries have shown.[48]

African Ethnicities and Slave Resistance in Sixteenth- and Seventeenth-Century Brazil

The regions of African slave supply for northeastern Brazil's booming sugar industry were tightly linked to the Atlantic geopolitics of the Portuguese slave trade. The two large African regions that the Portuguese controlled at least until 1650 were Greater Senegambia/Upper Guinea and West Central Africa.

In the early phases of Portuguese settlement in Brazil in the sixteenth century, most of the slaves arriving in Bahia and Pernambuco were Senegambians/Guineans, primarily of Wolof and Mandingo ethnicities, shipped through the Portuguese trading post of Cacheu. However, in the later part of the sixteenth century, slaves of various Bantu ethnicities arriving from West Central Africa began to replace the others in increasing number, until they formed the bulk of the enslaved workforce in seventeenth-century northeastern Brazil.[49]

The Portuguese had established relations with the West Central African kingdom of Kongo (whose king eventually converted to Christianity) since the 1480s, with a view to controlling the lucrative slave trade in the region. In 1568, as a result of the Portuguese-backed attack by the Jaga, the Kingdom of Kongo effectively became a Portuguese protectorate. In 1575, the Portuguese founded the port of Luanda in Angola, which became their most important center for the Atlantic slave trade and the shipment of slaves to northeastern Brazil, except during the period of Dutch occupation (1641 to 1648). In 1622, the Portuguese created the puppet state of the Kingdom of Ndongo. Eventually, the Portuguese were defeated and expelled by 1670, while the Kingdom of Kongo disintegrated and collapsed by 1689. However, the port of Luanda in Angola continued to supply Bahia and Pernambuco with the majority of their African slaves.[50]

Also as a result of the large number of African-born slaves mostly coming from the same region of West Central Africa, already from the beginning of the seventeenth century, large-scale actions of resistance were common in northeastern Brazil. They resulted mostly in the creation of communities of runaways (*quilombos* or *mocambos*), which, also aided by the presence of a still largely unexplored frontier area, succeeded in remaining autonomous for longer than anywhere else in the Americas.[51]

One of the earliest and largest of these runaway communities was Palmares, in the captaincy of Pernambuco. Lasting from the early years of the seventeenth century until 1694, despite repeated attacks by both Portuguese and Dutch authorities, Palmares was a large community of runaway slaves organized in different settlements and reaching a population of 11,000. In its later years, Palmares had a "king," Ganga Zumba, who attempted to negotiate with the Portuguese and was overthrown and killed by his nephew Zumbi. The nephew became the last king, and was eventually captured and executed after Palmares's fall.[52]

Besides its importance in the context of studies of slave resistance, Palmares is equally important for understanding the extent of the slaves' retention of African cultural traits and of their adaptation to the New World. A number of features in Palmares—from the tradition of kingship and royal lineage to the practices of agriculture, trading, and even raiding, down to the way the villages were built and enclosed in palisades—point to African legacies. Ultimately, as Stuart Schwartz has written, "while Palmares combined a number of African cultural traditions and included among its inhabitants *crioulos* [Brazilian-born blacks], mulattoes, Indians, and even some renegade whites, or *mestiçoes,* as well as Africans, clearly the traditions of Angola predominated." Therefore, the specific features of Palmares must be related to the legacy of West Central African traditions, and their study could enlighten us on the slaves' role in the early presence of African culture in the Americas.[53]

African Ethnicities and Slave Resistance in the Early British Caribbean

In the seventeenth century, the English traded slaves from Africa to their Caribbean sugar islands, first mainly from the Gold Coast and then also from the Bight of Biafra. Consequently, in the sugar plantations of British Caribbean islands such as Barbados and Jamaica, many slaves belonged to ethnicities such as the Akan/Aja from the Gold Coast, also known, together with the other Gold Coast ethnicities, as *Coromantee* or *Coromantins.* Only later, slaves of ethnicities such as the Igbo were traded from the Bight of Biafra.[54]

Therefore, at the beginning of the process of transformation of Barbados and Jamaica into slave societies, West African identity was still very strong, given that most slaves came from related ethnic groups that had lived in contiguous areas and spoken mutually understandable languages in Africa. David Eltis has gone as far as claiming that "seventeenth-century Barbados and Jamaica thus had two core cultures, the one, southern English, the other, Akan/Aja." Later evidence from slave cultural practices and also from slave resistance seems to support this claim, not just for Barbados and Jamaica, but also for the other areas of the British Caribbean during this period. Indeed, Trevor Burnard reports that "a major conspiracy uncovered in Antigua in 1736, for example, revealed a Coromantee leadership that sought to replace white authority with a Coromantee kingdom." Some historians, however, have pointed out that, aside from this dominant ethnicity, other regions of Africa—such as West Central Africa and even Madagascar—with different ethnic groups, even though in smaller numbers, were represented as well by slaves living in the British Caribbean islands in the seventeenth century.[55]

In Barbados at the beginning of the 1660s, with the continuous arrival of an increasingly larger number of African slaves bound to work in the fast-growing sugar industry, the ratio of whites to blacks began to shift dangerously, causing a great deal of anxiety among the white population and the authorities. Consequently, a strict slave code—which pointed out, together with the slaves' racial inferiority and property status, their propensity to rebel and the necessity of harsh punishments—was devised in 1661, followed by a more elaborate version in 1688. This became the model for all the slave codes enacted in the British Caribbean—first and foremost Jamaica, whose 1696 slave code made similar points to the Barbadian one in regard to slave rebelliousness and consequent exemplar punishment.[56]

Fears of slave rebellion were hardly unfounded, since the harsh working regime and brutal discipline that characterized sugar plantations combined to create the ideal conditions for uprising, especially taking into account also the slaves' close ethnic and linguistic relations. The authorities on Barbados discovered conspiracies involving large numbers of slaves on a number of plantations in 1675, and again in 1692. Significantly, in both cases, according to the documents, Gold Coast Akan/Aja slaves, or Coromantins, played a leading role.[57]

African Ethnicities and the Making of Slavery in Virginia

Most of the African slaves that arrived in Virginia in the seventeenth and eighteenth centuries were from Greater Senegambia/Upper Guinea, mainly of Mandingo and Wolof ethnicities. In the Chesapeake region, however, a large number of Igbo slaves came from the Bight of Biafra. The first slaves imported directly from Africa appeared in Virginia as early as 1619. The episode is famously documented in a letter in which Virginian planter John Rolfe described the sale of "20 and odd Negroes" by a "Dutch man of Warr." The letter does not give any additional details, but it is very likely that the Dutch ship cited in the document was involved in acts of piracy against the Spanish-Portuguese empire, and that the slaves might have originally been destined for the Spanish colonies.[58]

During the following 60 years, a continuously increasing number of slaves reached Virginia through the Atlantic slave trade. For the most part, though, the slaves that arrived in the Chesapeake Bay did not come directly from Africa, but from the Caribbean, and they were *seasoned.* In the Caribbean, the African slaves had the time and chance to adapt to the New World, especially to the new diseases, and to learn the rudiments of the English language, before being sold to the plantations of the North American colonies.[59]

Later on, after 1680, Virginian planters abandoned the use of seasoned slaves and preferred to acquire slaves that came directly from Africa. The reasons were many, among them the facts that the latter were cheaper, and also, not knowing the environment or the language, that they were less prone to stage rebellious acts.[60]

From the 1660s onward, with a series of particular laws called *slave codes,* Virginia's colonial government established that, unlike indentured servitude, slavery was a permanent condition, which lasted for life and was passed on to one's descendants, and radically limited the rights of both African slaves and free Africans. The colonial government also dissuaded masters from manumitting their slaves and sanctioned the full legality of heavy corporal punishment for rebellious slaves. These laws were followed by even more restrictive measures at the beginning of the eighteenth century.[61]

African Ethnicity and Slave Resistance in Early Colonial South Carolina

Very different from Virginia was the case of South Carolina, where, from the beginning, a large workforce formed exclusively of either African or Caribbean slaves toiled in the rice plantations.

The African slaves who arrived in South Carolina were also from Greater Senegambia/Upper Guinea and of mostly Wolof and Mandingo ethnicities. The slaves of African descent proved the only ones who were able to work in the particularly unhealthy environment of the coastal swamps, which was plagued by malaria and mosquitoes, and fled by the South Carolinian planters themselves for half a year, every summer.[62]

In South Carolina, the presence of an ever-increasing number of slaves from Africa became the main cause for the promulgation, starting from 1690, of special laws that, similar to the ones approved in Virginia, restricted radically the freedom of all colored persons and decreed the permanent, lifetime, status of African slavery. Other laws, even more restrictive, followed in 1696, 1712, and 1742. The progression of these dates, which coincides with the succeeding steps in the rise of rice production stimulated by the world-market demand, shows how the increasing legislative rigidity of the South Carolina slave system served fundamentally to guarantee the optimal conditions for the planters' use of the enslaved African workforce on their rice plantations, comparable to what happened in Virginia with the tobacco planters.[63]

However, in South Carolina there was also another reason for the promulgation of particularly restrictive laws toward all the population of African descent: already a short time after its foundation, the colony was characterized by a black majority. As a consequence, white South Carolinians always feared the possibility of a massive rebellion of all the colored people—slave and free—against them. The consequent restriction of basic rights hit the free colored population of South Carolina earlier and harder than in other areas of the American South.[64]

Starting from 1720, the year when slave import from Greater Senegambia/Upper Guinea reached 1,000 per year, together with the special laws already mentioned, new provisions, including the institution of special "black patrols," were added in order to control as thoroughly as possible the activities of the black community, especially in Charleston. And to be sure, a number of plots and attempted insurrections were discovered in South Carolina during the colonial period, the most famous of which was the Stono Rebellion of 1739.[65]

For the previous ten years of 1730 to 1739, planters imported large numbers of Angolan slaves from West Central Africa. In 1739, those Kongo slaves imported from Angola gathered at the Stono river and played a leading role in starting a rebellion, which occurred on a relatively large scale and led to the murder of more than 30 whites. Eventually, the local white militias managed to counterattack and defeat the rebels.[66]

The impression the Stono Rebellion left on whites was big enough to lead planters not just to abandon West Central Africa and revert to imports from Greater Senegambia/Upper Guinea as a source of slave supply, but also to enact even more restrictive laws. These new laws denied rights such as the use of weapons, and even the liberty to travel from one place to another, to all colored persons, both free and slave.

In his overview of the first two centuries of slavery in North America entitled *Many Thousands Gone* (1998), Ira Berlin used the expressions *charter generations* and *plantation generations* in reference to the earliest experiences of slavery by Africans in the American South. The former refers to "the first arrivals, their children, and in some cases their grandchildren." The latter refers to those "who were forced to grow the great staples." It is useful to apply these two definitions to all the first generations of African slaves in the plantation areas of the Americas in the seventeenth and eighteenth centuries. The early experiences of these first arrivals and forced settlers through the Atlantic slave trade had much in common in terms of work, oppression, legal control, and cultural adaptation. They lived in regions that underwent, at different times and as a result of different factors, the transformation from societies with slaves to slave societies.[67]

In his study, Berlin has termed the representatives of the charter generations as *Atlantic creoles,* who "might bear the features of Africa, Europe, and the Americas in whole or part." In fact, the clustering of ethnicities in particular regions of the New World allowed the retention of many features of particular African cultures that played major roles especially in inspiring specific acts of resistance and rebellion among the slaves. By the time of the plantation generations, and even more so of the subsequent *revolutionary generations,* the process of creolization had proceeded further. By then, retention of African traits had combined with the creation of a new syncretistic form of distinctively African American culture in all the major slave societies of the New World.[68]

The Making of an Atlantic Slave System

By the seventeenth century, a fully developed Atlantic slave system had come into being. Broad comparative studies by Immanuel Wallerstein and Eric Wolf have shown how the Atlantic slave system was a central feature of a European-dominated world economy. From the very beginning of the early modern era, this economy took advantage of peripheries such as the American colonies—and equally so of eastern and southern Europe, Africa, and Asia—by imposing on them highly exploitative labor regimes for the production of different types of raw materials. In eastern Europe, the dominant crop was grain. In the Americas, by the seventeenth century, the mechanisms of the European-dominated world market had led to a spectacular rise in the importance of sugar, which, until the beginning of the nineteenth century, continued to be the main product of a number of European colonies in the Americas.[69]

Both implicit, or *soft,* comparative studies and broader, explicit, comparisons have shown how the truly epochal and novel feature for the future of the Atlantic slave system was the Portuguese focus on sugar plantations. Adapting techniques already employed by Venetian and Genoese merchants in the medieval Mediterranean, the Portuguese made heavy use of African slaves in northeastern Brazil beginning in the 1570s. Also crucial was the later adoption of sugar plantation and African slavery first by the Dutch, and then by the French and the English. Mainly thanks to sugar, already by the seventeenth century, the Atlantic slave system included several fully developed slave societies, mirroring the unfree labor system that had sprung up everywhere in eastern Europe with the rise of the *second serfdom.*

Comparative history has shown clearly how every crop produced in other regions of the Americas, including Virginian tobacco and South Carolinian rice, was of secondary importance compared to Brazilian and Caribbean sugar. Thus, in comparative perspective, the British colonies in North America were at the periphery of the Atlantic slave system. Yet, they were very much part of the larger British Atlantic network—which included the Caribbean islands, and first and foremost Barbados and Jamaica—given the ties between the master classes, and even more important, their extended commercial exchanges. In this respect, Atlantic historians, by focusing on the early modern British Empire, have done much to help us understand the main features of a shared northern European culture among the master classes of the different British slave colonies.[70]

This was a culture in which aggressive Protestantism, obsession with racial purity and superiority, and mostly unrestrained merchant capitalism went hand in hand. And this behavior was mostly in opposition to comparable but competing models of master-class culture prevalent in Iberian and French colonies, as Robin Blackburn's important analysis of comparative links between the colonial Americas master classes and their European antecedents has shown. At the same time, though, Rafael Marquese's important work has shown in comparative perspective the presence of shared elements of modernity among the eighteenth-century slave societies of North America, the Caribbean, and Latin America.[71]

In regard to the slaves, equally important and innovative studies have shown us how, for those Africans forcibly transported on the different middle passages of the Atlantic slave trades managed by European powers such as the Portuguese, Spanish, Dutch, French, and English, the seventeenth and eighteenth centuries were crucial times of cultural transformation and adaptation. The slaves experienced processes that contained within themselves the seeds of another type of Atlantic modernity, alternative to the one advanced by the planter elites.[72]

Wherever they were taken in the New World, African slaves coped with the brutality, oppression, and legal exploitation brought by the plantation system. However, they also engaged in processes that led them, on one hand, to retain many of the cultural traits of their African regions of origin—particularly in regions strongly characterized by clusters of ethnicities—and on the other hand, to begin a process of creolization, which in time would create distinctive and syncretistic African American cultures in all the major slave societies of the Americas.[73]

In a wider context, it is important to remember that both broader and more specific comparative studies have done much to show us in detail how the rise of New World slavery was contemporaneous to the rise of the second serfdom in early modern eastern Europe. Grain became the main staple crop produced by Polish, Hungarian, and Russian masters all over eastern Europe at the same time, the late 1500s, that the Portuguese were making their early experiments in sugar production and African slavery. Both the Atlantic slave system and the eastern European serf system were systems of unfree labor that centered on production of crops in large landed estates—the plantations and the *latifundia.* In both systems, powerful landed elites ran large-scale agricultural enterprises, either partly or mostly oriented toward a world market.[74]

Thus, by some accounts, the origins of the modern features of New World slavery ultimately relate to epochal changes that, through the rise of an integrated capitalist world market, led either to the creation or to the reappearance of more or less modern systems of unfree labor. And even though historians have proven little direct and exclusive connection between the profits from the Atlantic slave trade or the plantation system and the capital that financed the Industrial Revolution (contrary to what Eric Williams had argued), a systematic comparative analysis of the rise of systems of unfree labor in the Americas and in eastern Europe in the early modern era would help us understand better also the deeper global implications of the making of free and also nominally free wage-labor systems in the modern West.[75]

Chapter Two

COLONIAL SLAVE SOCIETIES BETWEEN REFORMS AND REVOLUTIONS

During the eighteenth century, colonial slave societies in the Atlantic world reached their maturity. The plantation economy, which had taken off in the previous century, reached very high levels of production in the slave-based agricultures of the Americas' mainland and the Caribbean, as a result of the rising world-market demand for crops such as sugar, tobacco, and rice. As a consequence, planters in the French and British Caribbean, British North America, and Portuguese Brazil increased their wealth and power further. At the same time, the Atlantic slave trade guaranteed a continuous import of slaves for the New World plantation economy.[1]

In general, for slaves throughout the Americas, the eighteenth century represented a transitional period between two phases of expansion of the slave economy. It was a time that saw the first generations of individuals of African descent born and raised in the Americas, and one that, thanks to the slaves' still strong contacts with the African origins, saw also the beginnings of syncretistic African American cultures. For the planters, the eighteenth century represented a particularly important period in the process of ideological formation of master classes whose habits and lifestyles became characterized increasingly by the combination of European aristocratic cultural models with New World entrepreneurial attitudes.

Toward the end of the eighteenth century, *cultural transfers*—the processes of transmission of cultural models, according to the practitioners of "transfer history"—became increasingly frequent among the slaveholding elites in the New World and beyond. As a consequence, it is possible to see how, at this time, American slaveholders participated in an elite reformist culture, spread throughout the Atlantic, which attempted a rationalization of both agricultural production and slave management, as a result of different influences.[2]

Above all, this was an age of Atlantic revolutions. To the reformist and relatively conservative revolutions of the slaveholding elites in the United States and in Latin America, the slaves responded with the ultra-radical revolution that stemmed directly from events in France, and eventually created the Haitian Republic in the French Caribbean colony of St. Domingue. Here also, the notion of cultural transfers helps in understanding how the slaveholding elites throughout the New World coped with the subsequent economic disruption and ultimate adjustment of the Atlantic slave system in revolutionary times. Additionally, it helps us comprehend how the slaves, in different but related ways if we compare the American and Haitian revolutions, took to their radical and extreme consequences both the revolutionary ideologies and the opportunities that the revolutionary events offered them.[3]

MATURE COLONIAL SLAVE SOCIETIES IN THE ATLANTIC WORLD

In *Many Thousands Gone* (1998), Ira Berlin referred to a "plantation revolution" in describing the complex economic changes caused by the affirmation of the slave system in the American South between the seventeenth and the eighteenth centuries. This is a useful concept, which we can extend and apply to comparable economic transformations underwent by all the major commodity frontiers of the Atlantic world characterized by a plantation economy around the same time. In all of these frontiers, planters played a leading role in the making of the plantation revolution, promoting an ever-larger production of those crops—especially sugar, tobacco, and rice, which were more sought after by the eighteenth-century world-market demand.[4]

The profit made with the sale of those crops allowed planters to confirm once and for all their status among the most privileged classes in the European colonies and to dominate societies whose pyramidal structures were characterized everywhere by the presence of an increasingly larger population of people of African descent. The plantation revolution had crucial repercussions on the societies of the American colonies in the Atlantic world, and its consequences conditioned their subsequent developments.

Essentially, the plantation revolution led to the further transformation of the most productive agricultural regions of the Atlantic world into mature and full-fledged *slave societies,* according to a terminology first used by ancient historians Moses Finley and Keith Hopkins, and then also adopted by Ira Berlin. Essentially, in mature colonial slave societies—as Virginia, South Carolina, Barbados, and Brazil had become in the course of the seventeenth century—slavery was a central part of the economy. Planters were a very large elite, and those who aspired to become slaveholders formed the majority of the population.[5]

In every plantation economy of the Atlantic world, it was the planters' decision to focus exclusively on the cultivation of particularly valuable crops that conditioned the development of particular socioeconomic systems—systems whose survival depended on the continuous use of slave labor, mostly imported from Africa following the massively rising volume of the Atlantic slave trade. As a consequence, in all the slave societies in the Americas, continuing a pattern that had begun in the mid-seventeenth century, the plantation economy was firmly in the planters' hands, while social organization obeyed hierarchies based on the number of slaves that different individuals owned.[6]

Masters and Slaves in the Eighteenth-Century American South

In the slave societies of the British colonies in the American South, planters were particularly conscious that being at the head of a hierarchy based on slaveholding, rather than landowning, distinguished them deeply from Europe's aristocracy. For this reason, they attempted to compensate for this difference by imitating as much as possible the habits and ways of life of the English gentry. Conversely, other features, such as the fact that their social position was tied to the market demand for certain products and also their definite entrepreneurial attitude, made planters perhaps more similar to an elite group of merchant patricians of a sort. During the eighteenth century, the hybrid characteristics of the American planter elite—which effectively combined together, in almost equal measure, features related to both the Old World and New World—became particularly evident. Planters increasingly became a self-conscious class of privileged individuals and their families, whose generational ties related to the Americas, rather than Europe.[7]

In Virginia, where tobacco was the main cause of the plantation revolution, the increasing world-market demand led to the planters' accumulation of fortunes comparable to those of Europe's elites. For example, in 1728, Robert "King" Carter owned almost 400 slaves distributed in 48 landholdings.[8]

Given that the reputation and fortunes of Virginia planters depended on their ability to grow a highly marketable quality of tobacco in very large quantities, it is easy to see how this business required a marked entrepreneurial attitude, as all the documents of the time attest. And yet, the fact that a lot of the capital accumulated was spent in conspicuous consumption for the building of ever-larger and more imposing "Big Houses" tells us about the fundamentally hybrid character of the Virginian tobacco planter elite.[9]

In the case of the rice planters of eighteenth-century South Carolina, the hybrid character showed in other ways, since the plantation revolution had given origin to a unique slave system in the Americas. Only on the coastal swamps of South Carolina, during the summer season, did rice planters flee their plantations and their unhealthy climate to live in their town houses in

Charleston. Effectively, for half the year, South Carolinian rice planters left plantation management in the hands of their overseers, behaving in a comparable way to European *rentier* aristocrats, who were notoriously absentee landowners. At the same time, South Carolinian planters were particularly careful in conducting the business of rice cultivation from the distance, not giving up their entrepreneurial vocation. Therefore, also in this case, even though in a different way, we can talk of hybrid characteristics of the American South's slaveholding elite.[10]

For the slaves, the transformations related to the plantation revolution in Virginia and South Carolina had incalculable consequences. One reason was the increase in slave imports from Africa, resulting from the growing demand for workers in the tobacco and rice fields. Another was the worsening in terms of legal conditions and of sheer exploitation, as a result of the planters' attempts to make larger and larger profits from the sale of highly requested crops in the world market.[11]

However, the conditions of life of slaves in the American South were much better in comparison with the conditions of slaves in other parts of the Americas, such as the Caribbean and Brazil, where the mortality rate was appallingly high. This was especially the case on South Carolina rice plantations, where planters implemented the more humane *task system,* under which each slave needed to finish a certain task and then was free from obligations. This was opposed to the exhausting *gang system,* under which all slaves worked in gangs from dawn to dusk, in use on Virginian tobacco plantations.[12]

In general, in all the American South, the better working conditions, as well as the better diet and climate, produced a unique phenomenon in the New World with the constant rise in the rate of reproduction of the slave population. This, in turn, led to the relatively early formation of the first generation of *creole,* or American-born, slaves of African descent. As a result, uniquely among all the slave societies of the Atlantic world, in a relatively short time during the course of the eighteenth century, the total population of African Americans increased to reach around 700,000 by 1790, and then almost 900,000 only a decade later. In Virginia and South Carolina, the percentage of creole slaves ultimately succeeded in outnumbering the percentage of slaves imported from Africa. This important factor had crucial repercussions, especially on the development of the slave culture in the American South. It was at the origin of a process that John Blassingame has termed "the Africanization of the South and the Americanization of the slave." [13]

In practice, unlike what happened in other slave societies in the Americas, in the American South, the increasing outnumbering of African slaves by creole slaves led to a gradual loss, although one far from complete, of a number of cultural influences from Africa. The slaves' relatively rapid assimilation of American cultural traits led to the formation of an original and mildly syncretistic African American culture. This process was particularly evident in regard to religion. In this case, creole slaves assimilated traits of

Evangelical Christianity, to which they added those few elements of the African cultural patrimony that they still remembered. They created a unique southern tradition of African American worshipping—a process that began in the eighteenth century and reached its maturity in the American South in the following antebellum period.[14]

The British Caribbean and Eighteenth-Century Jamaica

In the British Caribbean, the eighteenth century saw Jamaica take the place of Barbados as the leading producer of sugar among the British colonies. By 1748, Jamaica produced three-quarters as much sugar as the other islands in the British West Indies. Similarly to what had happened in Barbados, the number of planters increased in Jamaica, together with the number of slaves imported from Africa to work on sugar plantations, leading to the making of a full-fledged slave society. Already by 1739, Jamaica had 429 sugar plantations and more than 100,000 slaves.[15]

Although there was much absenteeism that characterized the British planter class, a number of West Indian planters were resident slaveholders and landowners, who lived in luxury as a result of the profits they made in the capitalist world market through the business of sugar production. As Gad Heuman pointed out, "a medium-sized Jamaican plantation towards the end of the eighteenth century would have had at least 200 slaves and have consisted of approximately 900 acres." Similar to the planter classes of Virginia and South Carolina, Jamaican planters built large mansions, or Big Houses, at the heart of their plantations, thus providing another example of a lavish lifestyle, and also a comparable model in terms of hybrid characteristics. Jamaican society was at once entrepreneurial and capitalist-oriented, but also close to the idea of conspicuous consumption that characterized Europe's contemporary landed aristocracies.[16]

During the course of the eighteenth century, Jamaica imported a total of about 575,000 slaves from different regions of Africa, but mostly from the Gold Coast, and, later on, the Bight of Benin and West Central Africa. The overall living conditions, the relentless and back-breaking labor performed in gangs on sugar plantations, the inadequate diet, the appalling cruelty of overseers and planters (such as, famously, Thomas Thistlewood), and the widespread presence of diseases influenced demographics to such an extent that, by 1807, the total of Africans in Jamaica had decreased to 348,825 individuals.[17]

In time, as in the other mature colonial slave societies, the slaves in Jamaica who survived the hardship, together with the continuous new arrivals of enslaved Africans, became protagonists of a process of creolization of the slave culture. Similarly to what happened in Barbados, the Coromantee component and the Akan/Aja ethnicity from the Gold Coast were initially dominant, and then continued to provide a major factor of African identity.[18]

Specifically in regard to religion, in different regions of the British Caribbean, by the eighteenth century, African traditions, including the Akan/Aja, became slowly fused with elements of Christianity, giving origin to syncretistic religious traditions comparable to those of other New World slave societies. Clear signs of creolization and syncretism were present particularly in the West Indies African slaves' funeral practices. According to Philip Morgan, these funerals were characterized by "drumming, dance, and song; feasting and drinking, broken pottery, upturned bottles, and seashells marking black graves," and yet also "not far removed from the behavior of eighteenth-century evangelicals," and therefore from contemporaneous forms of Christianity.[19]

Slavery in Eighteenth-Century Brazil

Comparably to what happened in the American South and in the British Caribbean, during the course of the eighteenth century, in Brazil, the planter elite increased its hybrid characteristics, though in somewhat different ways. On one hand, the *senhores de engenho,* who dominated the sugar production on their plantations, behaved and were effectively treated—much more than American planters—as a colonial aristocracy of European descent, even though without any claim to hereditary titles. On the other hand, the fact that the wealth of the *senhores de engenho* was tightly related to the fortunes of the sugar industry and its performance in the world market "created," in Stuart Schwartz's words, "a highly volatile planter class, with *engenhos* changing hands constantly and many more failures than successes."[20]

As a consequence, Brazil's eighteenth-century planter class effectively ended up being an open type of elite, with the great slaveholding families who were third- to fifth-generation Brazilians not only intermarrying, but also allowing room for newly arrived Portuguese merchants and immigrants to replenish and renew the elite's ranks. Contrary to what older studies used to claim, most *senhores de engenho* were resident landowners and slaveholders who lived in their *Casa Grande* (Big House) and also had urban residences in the port cities of Salvador da Bahia and Recife, in Pernambuco, where they conducted a number of business activities, akin to the master classes of all the mature slave societies in the colonial Atlantic world.[21]

During the course of the eighteenth century, Brazil's northeastern sugar economy went through a major crisis, as neither Bahia nor Pernambuco managed to keep pace with the new areas of sugar production in the New World. Specifically, first Jamaica, and then St. Domingue rapidly outpaced Brazilian planters and replaced Brazil as world centers of sugar production. Still, during the eighteenth century, sugar accounted for one-third of Brazil's total exports. After decades of deepest crisis, the reforms promoted by the enlightened Portuguese government led by the Marquis de Pombal in the 1750s and 1760s succeeded in partly reviving the sugar economy of Pernambuco and Bahia, but Brazilian sugar no longer enjoyed the same prominent role it had in earlier times in both the internal and external markets.[22]

In fact, for the most part of the eighteenth century, the leading sector in Brazil's colonial economy was gold mining. Gold was first discovered in Minas Gerais in 1689 and 1690, and then in Goias in the 1720s and in Mato Grosso in the 1730s. Minas Gerais rapidly became the center of Brazil's gold-mining boom. As an increasing number of slaves were shipped to work in the mines, a new and different type of slaveholding economy developed. Effectively, slaveowners—particularly the ones who owned 12 or more slaves—were the only individuals to whom the Portuguese Crown granted mining concessions. By the 1760s, Minas Gerais had a population of 71,000 whites and 249,000 blacks, both slave and free. But note that the rate of *manumission*—the master's voluntary act of freeing a slave—was higher here than in any other Brazilian province. As a consequence, by 1808, the free colored population became more numerous than the enslaved colored population.[23]

Brazil's gold-mining economy maintained its prominence until the end of the eighteenth century. Then agriculture again became the leading entry in Brazil's exports, with the coffee boom in the south of the country. As scholars have pointed out, urban life thrived in Minas Gerais and the surrounding areas as a result of the mining boom. The province itself had a number of regional urban centers, starting from Ouro Preto. Another factor was its proximity to Rio de Janeiro (only 200 miles away), which led to the rise in size and importance of the southern Brazilian city as a slave market and a trading port.[24]

In the eighteenth century, the process of creolization reached new heights in the Brazilian cities. The organization of the first *Irmandades* (Brotherhoods) along color lines provided a new sense of New World identity for both slaves and free blacks whose ethnicities were distributed all across their different West African and West Central African regions of origin. As in later times, the *Irmandades* functioned both as centers of social activity and of religious celebration. An example is the *Nossa Senhora do Rosario* (Our Lady of the Rosary), which was present in all the major towns in Minas Gerais, starting from the 1720s. The *Irmandades* provided a milieu not just for the congregation of the enslaved and free black populations, but also for the elaboration and development of a range of new types of syncretistic Afro-Brazilian cultural activities.[25]

In comparative perspective, it is particularly striking to see how, in all the "many souths" that formed the mature New World slave societies of the Atlantic World, the planter elites shared similar types of ideology and behavior. In all the North American, Brazilian, and British Caribbean commodity frontiers with major areas of production centered on slaveholding and plantation management, the planter elites treated the cultivation of crops—whether sugar, tobacco, or rice—as a true business from which to make as high a profit as possible. This would certainly indicate that all of the elites behaved akin to real capitalist entrepreneurs.

At the same time, though, we also know from a number of studies that, even though in different ways and settings, all the successful planter elites of the New World engaged in one form or another of conspicuous consumption. They built very large mansions with refined furniture, and often even dedicated themselves to pastimes with features that made them somewhat akin to the European landed aristocracies, thus combining Old World and New World traits in a hybrid lifestyle.

To a certain extent and in somewhat different ways, slaves mirrored the planters' blending of Old World and New World traits by retaining a great deal of features related to their African roots in all the mature colonial slave societies where they were forcibly transported, at a time when the Atlantic slave trade reached its peak. Both West African and West Central African slaves—no matter how scattered they found themselves in the commodity frontiers of the American South, Brazil, and the British Caribbean—continued to maintain a connection with the traditions of their African regions of origin.

At the same time, the existence of increasingly larger numbers of slaves and also of free blacks and mulattoes born in the Americas (numbers that varied enormously from area to area) provided a new and different dimension to the African cultures in the New World. The result was an acceleration in the process of creolization of the plantation generations of African slaves and slaves of African descent in the Americas. Eventually, the outcome was the making of novel, syncretistic, cultural traditions, both African and American, though with different degrees of presence of the two elements, in all the major slave societies in the New World.[26]

REFORMING THE ATLANTIC SLAVE SYSTEM

Among the New World slaveholding elites in the second half of the eighteenth century, a similar culture of reform—prompted by Enlightenment thought, religious revival, and fear of radicalism—informed the actions of a number of planters in regard to management of both their land and workforce. More efficient systems of production, sometimes accompanied by more contractual forms of rule and based less on simply brutal force, were becoming increasingly popular, even before the Age of Revolutions (1770s to 1820s) began on both sides of the Atlantic world.

In the Americas, this culture of reform was particularly common among those planters who wished to ameliorate slavery by making it an even more efficient and profitable system, but maintaining, at the same time, an ambiguous relationship *vis-à-vis* their full involvement in capitalist economic and social relations.[27] For this reason, many slaveholders turned to the employment, in different degrees, of more paternalistic forms of rule over their enslaved workforce. They used forms of rule through which they could obtain higher profits and keep a low level of social conflict, tying even more tightly their crop production to the world market, while at the same time not changing the fundamental features of their unfree labor system.[28]

Comparison shows some interesting common patterns among the different planter elites' attempts at reforming the mature Atlantic slave societies. In practice, in all the major plantation economies in the Atlantic world, planters attempted to reform the slave system by relying on modern writings in political economy, by turning to the insights coming from the reading and contemporary interpretation of classical authors, and by attempting to improve their own profile in relation to moral and religious values. Therefore, it is fair to say that, also in their reform attempts, New World planters showed a somewhat hybrid attitude that combined at once entrepreneurial and aristocratic characteristics.

Planter Ideology and Culture in the Eighteenth-Century American South

In the British colonies in the American South, the richest and most influential members of the planter class imitated the lifestyle of the English gentry and considered their plantations the same as gentlemen's seats. The planters' preference for European cultural models was not confined to the architecture that characterized their Big Houses, but extended to education.[29]

While it is true that, in the eighteenth-century American South, most members of the planter elite had private tutors who took care of their children's education, there were also prestigious educational institutions, such as William & Mary College (founded in 1693 in Virginia), where studies focused on the study of theology and classics, as in the best schools and universities in Europe. In the eighteenth century, the study of classical authors had become a fundamental part of the culture of the members of the planter elite in the American South. Documents such as the diaries of Thomas Jefferson and George Washington—both large tobacco planters living in Virginia—and also the types of volumes present in their libraries, attest to this interest.[30]

On one hand, an important reason for the attention to classical authors was the fact that American planters followed the cultural standards from Europe in regard to education, as noted earlier. At the same time, though, classical authors gave planters throughout the American South a model of behavior, since planters could actually see how their increasingly republican and antimonarchical ideals were not in contrast with slaveholding. Great Roman political figures had also struggled to cope with similar problems of crop-raising and slave management in their own large *latifundia,* which were, effectively, the ancient Roman equivalents of American plantations.[31]

During the course of the eighteenth century, for planters in the American South, the reading of classics in this sense accompanied the reading of up-to-date treatises in political economy. This was a new theoretical discipline whose rise, especially in England, was tied to the beginnings of modern industrial capitalism. Those treatises supported the view that slavery was a less efficient economic system than free labor, leading to the creation of an atmosphere conducive to reform and amelioration in terms of productivity, and therefore to a better performance of the plantation economy in the capitalist world market.[32]

A key factor in improving productivity lay in the treatment of the workforce, which for long had obeyed to the rules of a patriarchal ideal, according to which the master was to show his immutable and absolute authority over the slaves by keeping a harsh discipline through frequent punishments. However, starting from the 1740s, the influence of the religious reform movement called the Great Awakening affected deeply the behavior of southern planters. The movement's challenge to the established church's authority, together with the emphasis on redemption of sins in order to achieve salvation, led a number of slaveholders to change partly their attitude toward treatment of their slaves, placing more emphasis on the benevolent aspect of their authority.[33]

In time, planters succeeded in constructing a public image of themselves as benevolent patriarchs who took care of their "people"—as Virginian planter William Byrd II called all his dependents and slaves—and who behaved akin to "fathers" toward their "children." By doing this, planters effectively created the premises for a better atmosphere in the master-slave relationship, and thus one more conducive to easier work management and, consequently, higher levels of production.[34]

British Caribbean Planters in a Time of Change

In the British Caribbean, during the eighteenth century, as a result of the growing world-market demand, sugar production reached its highest levels. As a consequence, the West Indian islands' planter class grew in wealth and importance. However, this trend did not go unchallenged; competition by St. Domingue led to the latter eclipsing Jamaica's role as the world center of sugar production in the 50 plus years between 1740 and 1791. At the same time, as in the American South, the charge of unprofitability brought upon the planters' lifestyle, increasingly based on absenteeism, and upon the slave system as a whole by the new science of political economy forced the West Indian slaveholding elite to embrace a culture of reform and improvement in regard to both agricultural production and management of slave labor.[35]

In the latter part of the century, the rise in importance of the movement for the abolition of the slave trade gave origin to another powerful challenge, which would eventually lead to the end of British slave imports from Africa in 1807, and the following decline in importance of the British Caribbean islands as producers of commodities. However, in the period between 1791 and 1807, following the elimination of the commercial rivalry of St. Domingue (a consequence of the Haitian Revolution), British West Indian sugar production and exports increased again to meet the still growing world-market demand.[36]

Since the mid-eighteenth century, British Caribbean planters had experimented with improvement in agricultural techniques, through the use of new implements, different varieties of cane, and more efficient mills. All of this resulted in the amelioration, in terms of speed and efficiency, of the production of sugar on plantations as a whole. In turn, this attitude toward agricultural improvement went hand in hand with a different attitude toward slave management as well.[37]

As J. R. Ward pointed out, in the second half of the eighteenth century, British planters in the West Indies "began to show more concern for encouraging natural reproduction among their slaves, under the influence of the new humanitarian ideas from Europe and a marked rise during the 1760s in the prices charged for imported Africans." Slaves were better clothed and better fed. As the campaign for the abolition of the slave trade started, the "impetus to 'amelioration'" among West Indian planters reached its peak, "as planters tried to establish a self-sustaining labor force." However, in the long run, the planters' embracement of the culture of improvement and reform had little effect overall, since it allowed the British sugar-producing islands to remain reasonably profitable only until the British Atlantic slave trade was finally abolished in 1807.[38]

Eighteenth-Century Economy and Society in French St. Domingue

From the middle decades of the eighteenth century, the island of St. Domingue, the largest and most important in the French Caribbean, became the world center of sugar production. Despite a slow start since its acquisition by the French from the Spanish in 1697, St. Domingue's agricultural production, and consequently also its slave population, grew steadily, thanks to the ever-rising world-market demand for both sugar and coffee.[39]

On the eve of the French Revolution, in the 1780s, in the French Caribbean colony of St. Domingue, the population was composed of 40,000 whites, 28,000 free blacks and mulattoes (classified as individuals of free condition), and 452,000 slaves. Whites formed less than 10 percent of the population. Black slaves formed 87 percent of the entire population and were scattered among more than 8,000 plantations. The largest plantations were called *ateliers,* with an average size of 150 to 200 slaves.[40]

The reasons for the immense number of slaves lay in the continuously rising demand for both sugar and coffee by the European elites. This demand led St. Domingue, by 1789, to provide France with about two-fifths of its entire foreign trade and the world with half of the entire supply of both crops. As a consequence, comparably to what happened in other Atlantic slave societies, the import of a continuously higher number of slaves through the Atlantic slave trade reached its peak in the late eighteenth century. This was also the only effective means to replace the enormously high number of deaths that plagued St. Domingue's sugar plantations, as a result of the slaves' appalling living and working conditions, the cruelty inflicted on them, the diseases, and the inadequate diet.[41]

In an opposite pattern to the one of the increasingly larger number of African slaves, the number of whites in St. Domingue remained more or less the same throughout the eighteenth century. This, in turn, created a situation in which a very small elite group was made of large planters and merchants—the *grand blancs*—who owned enormous wealth, the majority of the land with the sugar and coffee plantations, and the majority of the slaves. Much less prosperous were the artisans, plantation managers, and shopkeepers, known as the *petit blancs,* who resented the power of the *grand blancs.* Similarly, the mulattoes, though free and in possession of about 100,000 slaves, resented the position of economic, social, and political inferiority in which they found themselves in comparison with whites, especially since Louis XIV's 1688 slavery legislation, the *Code Noir,* had provided for "free people of color ... to share the rights of other colonists."[42]

At the same time, it is also true that St. Domingue's large planters resented equally strongly the controlling policy of the French government, which regarded them as second-class citizens, even though the French economy needed badly the sugar produced on St. Domingue's plantations. The planters wished to be free to administer and decide the island's internal politics on their own.[43]

Thus, each major group of free and slaveholding individuals in St. Domingue had reasons to want major reforms. It was the French Revolution that provided whites with the opportunity of voicing their own wish for reform through a constitution limiting French control, drafted by St. Domingue's Colonial Assembly in 1790. It was the first step toward the unleashing of the Haitian Revolution.[44]

Altogether, the second half of the eighteenth century witnessed, on one hand, a major change in the cash-crop economy of the New World slave societies, as St. Domingue took its definitive place as leading producer of sugar in the Atlantic world. On the other hand, the influence of religious reform, Enlightenment thought, and classical capitalism gave crucial contributions to prompting changes in the slaveholders' ideological outlook. In different ways and degrees, all these factors posed challenges to the planter classes of the Americas. Slaveholders in both the British colonies in the American South and in the Caribbean responded to these challenges through an ideological adjustment that led them to embrace a culture of reform, or *amelioration.*[45]

At the same time, especially in the American South and in St. Domingue, as a consequence of a combination of different factors, planters became increasingly politicized and voiced their wish for reform as a form of discontent regarding their status as colonial subjects. Ultimately, in the British Caribbean, planters had little choice but to accept the growth of the movement for the abolition of the slave trade and the imposition of the latter's eventual end in 1807. Instead, in the American South and St. Domingue, this was the beginning of the process that would lead to the American and French revolutions, which allowed planters to take the initiative in the colonies' active political opposition to the metropolitan governments. However, unlike what happened in the American South, in St. Domingue, the planters' moment of independence did not last long, as the revolutionary initiative soon fell in the hands of first the mulattoes and then the slaves themselves.[46]

ATLANTIC REBELLIONS AND REVOLUTIONS

Ever since the publication of the now classic studies *The Problem of Slavery in the Age of Revolutions* (1975) by David Brion Davis and *From Rebellion to Revolutions* (1979) by Eugene Genovese, historians of slavery have been thinking comparatively about the Atlantic world during the Age of Revolutions—the period between the 1770s and the 1820s. Historians now focus increasingly on the crucial ideological links between the elites of the Americas and Europe, on the crucial influence that changes in the latter had on the transformations of the New World slave system as a whole, and on the repercussions of the French Revolution's radicalism on slave rebellions in an Atlantic context.[47]

At the same time, studies such as Robin Blackburn's *The Overthrow of Colonial Slavery* (1988) and his more recent *The American Crucible* (2011) have shown that it is impossible to understand the true significance of the end of colonial slave systems in the Americas without making constant references to the economic, political, and social situation in revolutionary and Napoleonic Europe. As a consequence, slavery scholars now see the American, French, and Haitian revolutions and their aftermaths as stages in a massive and interlinked series of upheavals that led to a crisis, and then to a readjustment of the New World slave system within an Atlantic context—a context properly understood as essential to a full and textured comprehension of the problems of slavery and freedom.[48]

From Europe, where they had originated in the works of philosophers and jurists, the founding ideas of the Enlightenment spread to the Americas, and through the publication of books and pamphlets, they had an incalculable influence on the master classes of the slave societies of the Atlantic world. This influence expressed itself particularly with the creation of a juridical culture whose fundamental principles condemned cruel and unjust practices. This culture attributed the highest importance to the individuals' natural rights, tolerance of differences, freedom of cult, and equality in front of the law.[49]

The influence of Enlightenment humanitarian doctrines reached its peak in the early phases of the French Revolution. Subsequently, though, and in a relatively short time, the antislavery momentum that characterized the Age of Revolutions winded up producing much more radical consequences—akin to the effects of the Jacobin period in revolutionary France—and led to a general crisis of the entire Atlantic slave system. The latter crisis occurred also as a consequence of the French revolutionary and Napoleonic wars and their influence on the economy of the Atlantic world. At the same time, the slaves' continuous unrest, which increased in the course of the eighteenth century, gave also an important contribution, leading to the making of large runaway slave communities and also producing some notable large-scale slave rebellions in the major slave societies of the Americas. However, the slave rebellion that occurred in French St. Domingue—then the richest sugar colony in the Americas—eclipsed all other episodes in terms of importance, leading to the Haitian Revolution, and with it to the epochal end of the colonial Atlantic slave system as a whole (see Map 2).[50]

American Slavery in the Revolutionary and Postrevolutionary Periods

More than in other slave societies in the Atlantic world, in the American South, a profound contradiction characterized the behavior of planters during the entire duration of the wars of the American Revolution (1775 to 1781). On one hand, many of them fought for the independence of the British colonies in North America. On the other hand, they denied freedom to the enslaved African American population. For their part, the British hit directly at the plantation system by issuing proclamations—first by Lord Dunmore in 1775, and then by Sir Henry Clinton in 1779—claiming that all the slaves who belonged to the rebel American colonists and joined the British Army would be freed.[51]

In practice, only about 1,000 slaves fled in order to join the British Army. Many more simply left their plantations, taking every opportunity they had to escape. In South Carolina alone, according to Philip Morgan, a quarter of the total slave population of the colony managed to flee from plantations and farms. Despite these devastations, South Carolina's rice economy was still strong even after the war, and soon rice planters sought to find a way to replace the slaves who had fled through the Atlantic slave trade.[52]

In Virginia and the surrounding areas, the war gave a hard blow to an already weak economic system, since from the mid-eighteenth century, the production of tobacco had entered a period of serious crisis. This led to an increase in cases of manumission, facilitated by the authorities with special legal measures. Eventually, while wheat replaced tobacco as the premier crop in a number of places, more and more slaves, especially those with special skills, tended to be hired out according to the need, rather than being kept in a single place.[53]

In the last two decades of the eighteenth century, uniquely among all the slave societies of the Atlantic world, as a result of the general crisis of the slave system, the newly born republic of the United States of America—freed from

British colonial rule since 1781—saw laws that facilitated manumission being enhanced in many states. These included Virginia, Maryland, and Delaware, where thousands of slaves were set free. The peak of this process came with a series of laws that, between 1780 and 1804, made slavery illegal in all the northern states.[54]

From the end of the eighteenth century on, the contrast between the northern states' urban and mercantile interests and the southern states' interests—still focused on slavery—became a major feature of American political life. The topic dominated debates within the constitutional convention, which, in 1787, resulted in the creation of the Constitution of the United States. In the course of those debates, the northern representatives obtained an important success with the abolition of the Atlantic slave trade fixed for 1808, a year after the 1807 abolition of the British Atlantic slave trade. However, the Constitution basically sanctioned the legality of the slave system and provided also for the support of the Federal Government of the United States in matters such as the recapture of fugitive slaves.[55]

Slave Resistance in Eighteenth-Century Jamaica

Even though not going through a revolution, throughout the course of the eighteenth century, the British Caribbean islands were in a continuous state of alert, as a result of routine episodes of slave resistance in many different guises. Particularly effective in this sense was the creation of Maroon communities by runaway slaves, akin to Brazil's *quilombos* and *mocambos,* of which the largest were found in the isolated mountainous regions of Jamaica.[56]

As Philip Morgan has remarked, "by 1739, when the colonial government of Jamaica recognized their free and separate existence [at the end of the First Maroon War], the Windward Maroons in the eastern mountains and the Leeward Maroons in the western interior had been waging war against whites for more than eighty years." Even though the 1739 treaties effectively sanctioned Maroon independence in exchange for peace and the Maroon promise to capture subsequent runaway slaves, further British attempts to take control of the two Maroon communities led to a Second Maroon War in 1795. The Leeward Maroons of Trelawny Town were defeated after two years of fighting and eventually deported to Sierra Leone.[57]

Aside from Maroon activities, Jamaica also witnessed several rebellions and also one of the largest scale episodes of slave revolt in the British Caribbean in 1760. The revolt began in St. Mary's Parish, when, on Easter night, about 100 slaves under the leadership of an Akan/Aja slave called Tacky rose and seized the local fort. Subsequently, the revolt spread throughout the island, involving perhaps 30,000 slaves overall, and even leading to the rebels' capture of the capital Kingston and the proclamation of a slave woman as the Queen of Kingston. Subsequently, in 1761, Jamaica's British colonial authorities managed to defeat

Map 2 The Age of Atlantic Revolutions and Slave Revolts (1770–1848)

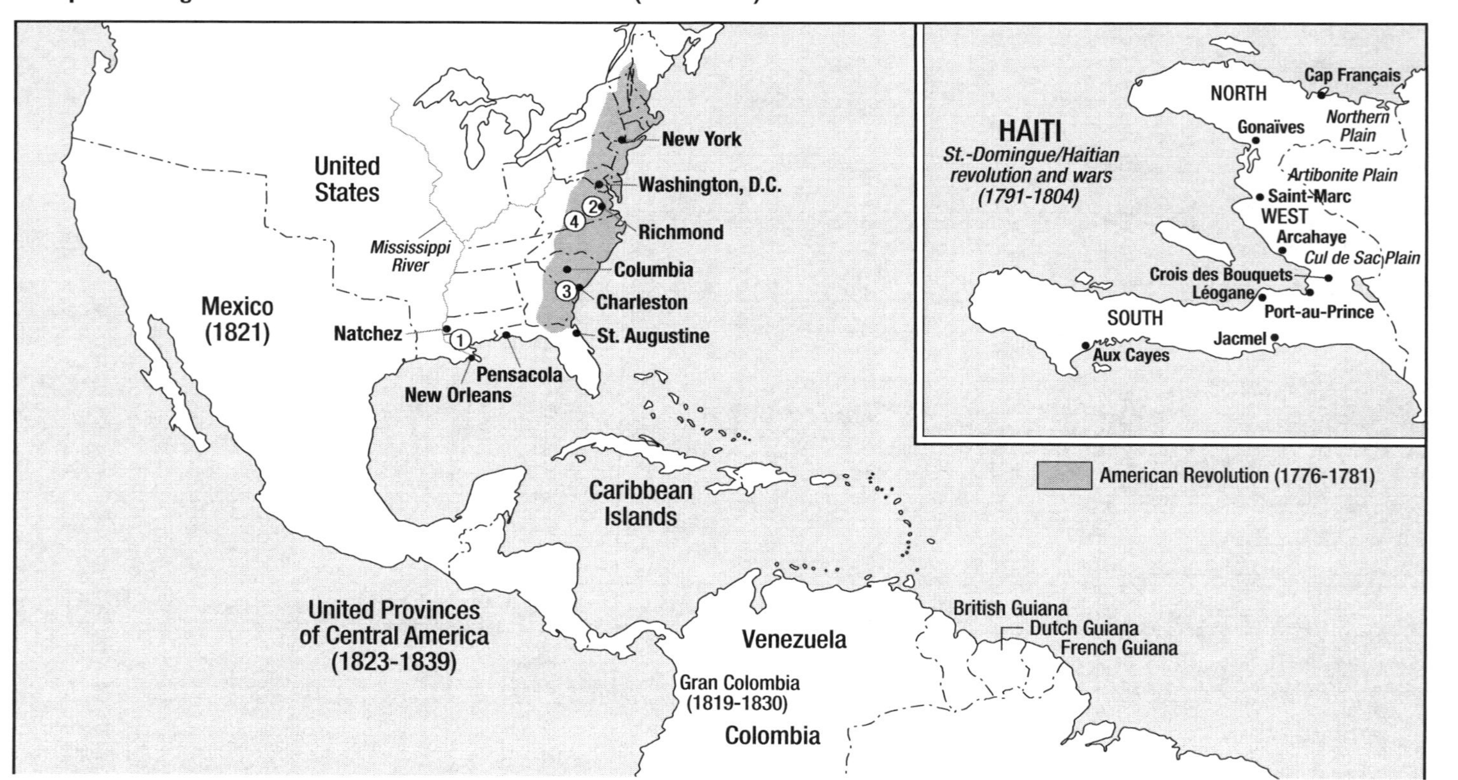

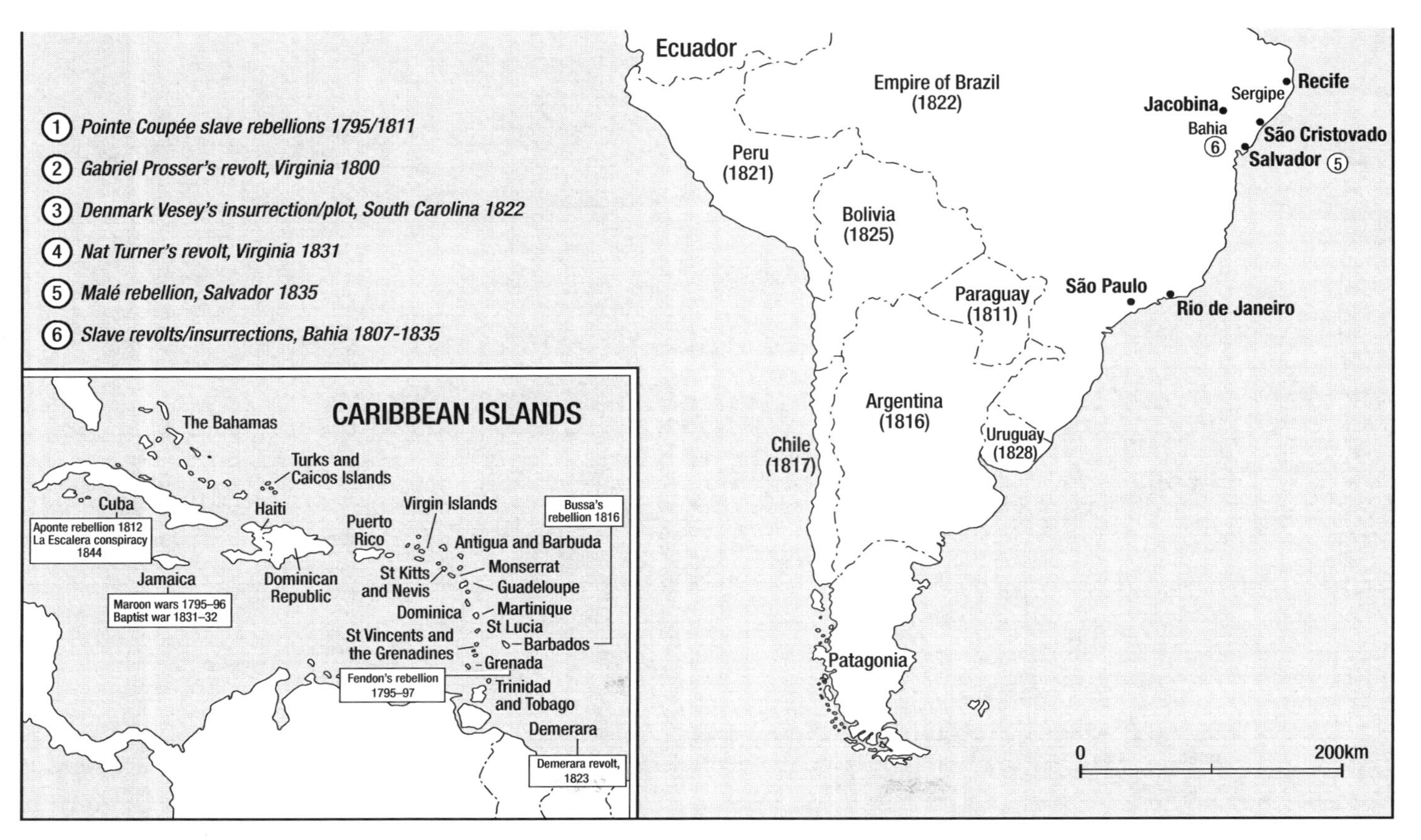
Ecuador
Empire of Brazil
(1822)
Peru
(1821)
Bolivia
(1825)
Paraguay
(1811)
Argentina
(1816)
Chile
(1817)
Uruguay
(1828)
Patagonia
Jacobina
Sergipe
Recife
Bahia
6
São Cristovado
Salvador 5
São Paulo
Rio de Janeiro
0
200km
1 Pointe Coupée slave rebellions 1795/1811
2 Gabriel Prosser's revolt, Virginia 1800
3 Denmark Vesey's insurrection/plot, South Carolina 1822
4 Nat Turner's revolt, Virginia 1831
5 Malé rebellion, Salvador 1835
6 Slave revolts/insurrections, Bahia 1807-1835
CARIBBEAN ISLANDS
The Bahamas
Turks and
Caicos Islands
Cuba
Aponte rebellion 1812
La Escalera conspiracy
1844
Haiti
Jamaica
Maroon wars 1795–96
Baptist war 1831–32
Dominican
Republic
Puerto
Rico
Virgin Islands
Bussa's
rebellion 1816
Antigua and Barbuda
St Kitts
and Nevis
Monserrat
Guadeloupe
Dominica
Martinique
St Lucia
St Vincents and
the Grenadines
Barbados
Grenada
Fendon's rebellion
1795–97
Trinidad
and Tobago
Demerara
Demerara revolt,
1823

the rebels, but only after 90 whites and 400 slaves had been killed. After the revolt, 100 of the captured slaves were executed and 600 were exiled.[58]

Slave Resistance in Eighteenth-Century Brazil

In Brazil, as in the British Caribbean, the eighteenth century was a period of heightened unrest among the various slave communities. Both in the older areas of sugar production and in the newly settled regions where a slave-based economy took hold, the chances of slave rebellious acts were high, especially considering that the number of *quilombo* runaway settlements continued to increase even after the defeat of the Palmares rebels in 1695. In the course of the eighteenth century, *quilombos* became even more widespread. Together with Bahia, the area with the largest number of *quilombos* was the gold-mining boom province of Minas Gerais, with 160 of them. The largest of the *quilombos,* destroyed in 1746—the so-called Kingdom of Ambrosio, or *Quilombo Grande*—numbered close to 1,000 slaves living in several palisaded villages (*palenques*), according to Herbert Klein and Francisco Vidal Luna.[59]

In 1788, partly as a result of the crisis of the mining economy, Minas Gerais (specifically, Ouro Preto) became the center of a large conspiratorial plot (*inconfidencia mineira*), led by Joaquim José da Silva Xavier, known as *Tiradentes,* which aimed at proclaiming a Mineiro republic independent from the Portuguese government. Even though the plot was discovered and the conspirators killed, it is significant that in *Tiradentes'* mind, in the words of Dauril Alden, "the republic was to be democratically governed ... [and] would be defended by a citizen militia ... [of] Brazilian-born blacks and mulattoes, to whom the revolutionaries promised freedom." In perspective, the Mineiro conspiracy opened a period of 30 years in which a series of major slave revolts occurred in Rio de Janeiro (in 1794), in Bahia (in 1798), in Pernambuco (in 1801), then again in Bahia (in 1807), and again in Pernambuco (in 1817). These ended temporarily only when Portuguese King Joao VI's son Dom Pedro I proclaimed Brazil's independence from Portugal in 1822 and managed to assert his control and reestablish order over the entire country.[60]

The Haitian Revolution in St. Domingue

Compared to the American South, the British Caribbean, and Brazil, St. Domingue stands out as the only case in which revolution and slave rebellion combined to completely subvert a major mature colonial slave society and replace it with a black republic. Indeed, referring to Michel Hector and Laennec Hurbon's concept of Haiti's "triple revolution—against 'slavery, colonialism and racial oppression,'" Robin Blackburn has recently argued that the revolution "had a double consequence. It suppressed plantation slavery and affirmed racial equality." Therefore, it is no wonder that historians have been busy investigating the causes of the Haitian Revolution, given its status of exception among all the slave rebellions in the Atlantic world.[61]

Modern historical research has clarified that the causes of the Haitian Revolution in the French Caribbean colony of St. Domingue are related to a complex of social and demographic factors. A particular factor was the presence of a small number of free individuals, of whom only a fraction were privileged white planters, and of an overwhelming majority of black slaves. Another factor was the existence of a particularly harsh slave system, which had already created an explosive situation by the second half of the eighteenth century. It was the influence of the French Revolution in 1789, though, that provided the necessary catalyst for the overthrow of slavery in St. Domingue, also as a result of the Haitian leaders' adaptation of French revolutionary ideals, as both C. L. R. James' *The Black Jacobins* (1938) and Eugene Genovese's *From Rebellion to Revolution* had already demonstrated.[62]

Initially, during the early revolutionary period, the fact that St. Domingue's planters acted on their resentment toward the French government caused a first collapse of the sociopolitical system of the Caribbean colony. Exploiting the confusion and the problems that the French government went through during the early phases of the French Revolution, St. Domingue's planters declared independence from France. However, the mulattoes and free blacks also demanded acknowledgment of their rights. Therefore, a number of conflicts occurred between the two sections of the free population in an atmosphere of true civil war. It was at this point that the slaves saw they had an unmissable opportunity to overthrow their oppressors.[63]

Taking advantage of their superior numbers, in 1791, St. Domingue's slaves began a series of revolts. In the organization of these revolts, the networks established through the practice of the Afro-Caribbean Vodou cult—strongly influenced by the religion of the predominant Kongo slaves imported in large numbers from West Central Africa in the late eighteenth century, rather than the slaves from the Bight of Benin—played a crucial role. This earlier phase of the revolt led to the indiscriminate slaughter of both whites and mulattoes who lived in the northern part of the island. Subsequently, in 1792, the revolts extended also to the southern part of the island with the same result. Thus, in a relatively short time, slaves succeeded in taking control of the majority of the territory of St. Domingue, and by August 29, 1793, Toussaint L'Ouverture—effectively, the leader of the revolt—and Jacobin commissioner Léger Félicité Sonthonax were able to proclaim independently the end of slavery on two different areas of the island.[64]

An ex-slave who had proclaimed himself free, L'Ouverture was not only the rebellion's mastermind, but also a great military leader and tactician. Together with his newly raised army, L'Ouverture managed to defeat both the Spanish and the English, who attempted an invasion of St. Domingue between

1792 and 1793, and, later, various troops sent by the French revolutionary governments to reestablish French control on the Caribbean colony. As a result, already by 1794, St. Domingue was de facto acting with a large degree of independence. On February 4 of the same year, the French National Convention decreed the official end of slavery in all the French colonies.[65]

The subsequent rise of Napoleon Bonaparte led to the restoration of slavery in the colonies and to repeated attempts to reverse the situation in Haiti by France and also other European powers. In 1801, L'Ouverture repelled the last attempts at invasion by Spanish and English forces, together with another attempt to reestablish French colonial rule by Napoleon's General Charles-Victor-Emmanuel Leclerc, while he defeated once and for all the mulattoes' resistance in the island, and even managed to write a Constitution for St. Domingue. Even though, subsequently, the French managed to capture L'Ouverture and take him to France, where he was executed, resistance to Napoleon continued throughout 1802 and 1803, and eventually the Republic of Haiti succeeded in being recognized as an independent and sovereign state on January 1, 1804.[66]

Thus, uniquely in the entire Atlantic world, a slave rebellion had as its consequence the creation of a republic of ex-slaves who had proclaimed themselves free, as L'Ouverture himself had done first. Therefore, there is little wonder that Haiti's example continued to be a source of inspiration for all the slaves scattered in the Americas, whose highest dream was, if not to overthrow the entire slave system, certainly at least to flee to the existing haven for ex-slaves.[67]

A number of contemporary historical studies have shown how, thanks to the existence of Haiti as a living proof of success of the St. Domingue slave rebellion, the influence of the Haitian Revolution not only extended to the entire Atlantic world, but also continued long after the recognition of Haitian independence. On one hand, the collapse of the richest and largest sugar-producing economy in the Caribbean and one of the largest slave societies and plantation economies in the Americas was a major factor in the general crisis of the colonial slave economies at the end of the eighteenth century. On the other hand, that same collapse allowed first Jamaican and then Cuban planters—and in lesser measure, the planters of other Caribbean islands as well—to expand exponentially their sugar production, eventually taking the place left void by the end of St. Domingue's slave system as the premier sugar-based plantation economies in the Atlantic world for the best part of the nineteenth century.[68]

In comparative perspective, the period between the end of the eighteenth century and the beginning of the nineteenth century saw the remarkable rise of two opposed new republican societies in the Atlantic world: the United States, a republic that effectively protected the existence of slavery, and Haiti, a republic that had abolished slavery once and for all. At a time in which the Atlantic slave system was transforming as a result of the economic and political crises caused by the turmoil of the Age of Revolutions, the presence of these two countries with similar political organizations, but with opposite attitudes toward slavery is a highly significant feature. As much contemporary scholarship has established, in different ways and at different times, the American and Haitian republics—both unique political institutions in the Americas, at least until the creation of the Latin American republics out of the Spanish colonial empire in the 1810s and 1820s—influenced and conditioned, either directly or indirectly, the subsequent development of the neighboring countries, especially of the Atlantic slave societies in the New World, throughout the following century.[69]

The Atlantic World Toward the Second Slavery

In retrospect, it is possible to consider the conciliatory attitude of the United States Constitution toward slavery as representative of a particular climate, which toward the end of the eighteenth century was becoming again favorable to the slave system both in the American republic and in the Atlantic world at large. The main cause of this shift in opinion lay in the renewal of Atlantic slavery, which, having survived the major economic and political crises of the Age of Revolutions, became revitalized by a new cycle of expansion of the world economy. This expansion favored the world-market demand for both similar and different crops in comparison with the ones grown in the plantations of the mature colonial slave societies. In turn, the regions that specialized in the production of these crops in slave-based plantations in different areas of the Americas became the most successful slave societies and commodity frontiers in the nineteenth century.[70]

In order to distinguish this period of economic expansion of the Atlantic slave system, with its particular features, from the previous period of first expansion of the colonial slave economies, Dale Tomich and Michael Zeuske have used the term *second slavery.* This expression encompasses also a number of global changes that occurred as a result of the transformation of the Atlantic slave system.[71]

But the deep changes the Atlantic world experienced at the end of the eighteenth century in the general attitude toward slavery were not just a consequence of economic transformation. Throughout the regions with plantation economies in the New World, the conservative turn that characterized the post-revolutionary political elites also played a very important role. The same can be said in regard to the public opinion, since in the post-revolutionary period, fewer intellectuals advocated the ideals of freedom and equality that were the bases of the American and French revolutions. Instead, they embraced conservative ideologies focused on the defense of private property, in whatever shape that came, including slavery.[72]

The early 1800s were a time of general revival of the Atlantic slave system and of renewed general defense of the need for a slave-based economy, especially in those areas that profited from the economic changes related to the rise of the second slavery. At that time, the example of the free Republic of Haiti, whose citizens were self-proclaimed freed slaves, stood as, and remained for most of the nineteenth century, an immensely powerful reminder of the fragility of New World slave societies—economies and labor systems that continued to be based on the unjust exploitation of hundreds of thousands of individuals of African descent.[73]

Chapter Three

THE "COTTON KINGDOM," ITS NEIGHBORS, AND ITS CONTEMPORARIES

In the Americas, nineteenth-century agricultural production on large-scale plantations was tightly linked to the rise of the *second slavery*, a phenomenon that Dale Tomich and Michael Zeuske have described as "the formation of highly productive new zones of slave commodities in the Atlantic and particularly in the U.S. South, Cuba, and Brazil." In other words, in these areas during the nineteenth century, a process occurred—similar to the one Jason Moore has described in relation to early modern peripheries—through "which the production and distribution of *specific* commodities, and of primary products in particular . . . restructured geographic space at the margin of the [world] system in such a way as to require further expansion."[1]

Referring in particular to Immanuel Wallerstein's model of analysis of the historical expansion of the European "world-economy," Moore has talked of *commodity frontiers* in relation to the production of particular types of raw materials, including specific crops, and has argued that this "concept allows an exploration of the interrelationships between production in *one* place, and the expansion of capitalist *space* in general."[2]

THE SECOND SLAVERY AND THE ATLANTIC PLANTATION ZONE

The aggressively capitalist, profit-driven, world-market-oriented, and highly exploitative nature of the second slavery, though often camouflaged as paternalistic, is clearly recognizable in the organization of American, Cuban, and Brazilian plantations, and in their management of land and labor. Most of these plantations were completely geared toward production of particular staple crops that commanded the nineteenth-century world economy: cotton, coffee, and sugar. The slave societies present in these three regions of the Americas and their economies effectively represented the ultimate capitalist transformation of what Tomich and Zeuske have termed an *Atlantic plantation zone*, one whose commodity frontiers dated to the seventeenth century and had undergone a crucial renewal in the period between the eighteenth and nineteenth centuries.[3]

From the late sixteenth through early seventeenth century up to the last two decades of the eighteenth century—since the earliest efforts of England, France, and the Netherlands to colonize the New World—slavery in the Atlantic plantation zone had produced crops that had commanded the world market. During the colonial period, profitability in slavery was mainly associated with sugar production first in northeastern Brazil, and then in the Caribbean, particularly the French colony of St. Domingue, and with tobacco and rice production in the British colonies in the American South. Starting from the 1780s, a combination of factors led concurrently to a restructuring of the world economy, which, in turn was responsible for the rise of new crops to command the world market. This restructuring also contributed to the creation and subsequent expansion of new and increasingly larger areas of slave-based agriculture—areas in which expansion of production and of the capitalist space went hand in hand with environmental change and technological innovation.[4]

Among the factors that caused these monumental transformations, paramount were the 1791 Haitian Revolution, which deprived France and the world of the main producer of sugar and paved the way for other regions to take St. Domingue's place. Other factors were the proliferation of textile mills in England and, in lesser measure, in the northern United States, which caused a rising demand for cotton. There was also a growing demand for coffee as a primary luxury product. The result was a readjustment of the Atlantic plantation zone and the beginning of a new phase in its history.

In this new phase, a second slavery characterized the American hemisphere with markedly different features from the previous slave-production complex, while it expanded in "new zones of agricultural production outside of the colonial empires of France and Great Britain," in Tomich and Zeuske's words. Ultimately, the second slavery found its most complete expression in the plantations of three specific Atlantic regions: the U.S. South, Cuba, and southern Brazil.[5]

The Atlantic is crucial in the concept of second slavery, since the latter was fundamentally a phase or a period in the long span of Atlantic history. Lasting roughly 100 years starting from the 1780s, this period witnessed the concurrent decline of older regions of slave production and the expansion of newer regions such as the U.S. South, Cuba, and southern Brazil, all within the Atlantic plantation zone.

Yet, if we think on a truly Atlantic scale, we should not confine our studies to the Americas, since the slave systems of the U.S. South, Cuba, and Brazil under the second slavery were not the only novel features in nineteenth-century Atlantic slavery. In fact, an equally large-scale slave society—one of the biggest in the world in absolute terms—existed on the other side of the Atlantic Ocean, in West Africa: the Sokoto Caliphate.

THE SOKOTO CALIPHATE IN COMPARATIVE PERSPECTIVE

According to Paul Lovejoy, "it is even possible that there were as many slaves in the Sokoto Caliphate in the middle of the nineteenth century as in the United States at the outbreak of the Civil War." Therefore, there could have been 4 million slaves in the Sokoto Caliphate, corresponding to almost half of the population. In the United States, slaves made up one-third of the population of the South. Brazil had 1.5 million slaves, while there were fewer than 400,000 slaves in Cuba, respectively 15 percent and 27 percent of the total population of the two regions. Besides being a comparable large-scale slave society, the Sokoto Caliphate had also emerged at the beginning of the nineteenth century, at the same time as the renewal of the slave societies in the Americas under the second slavery.[6]

Despite the essential differences that existed between the Sokoto Caliphate and the slave societies in the Americas under the second slavery (chief among these differences was the pervasiveness of Islamic ideology on the entire Sokoto slave system), there are important comparable points that are fruitful to explore. These will help us to better understand, through the analysis of both similar and different features, what really made slavery in the U.S. South, Cuba, and Brazil advanced in capitalist terms and unique in ideological and cultural terms. These features of the slave societies in the Americas under the second slavery become particularly clear when compared with the features that characterized an equally large-scale, but much less capitalist-oriented nineteenth-century slave society, such as the Sokoto Caliphate.

In the Sokoto Caliphate, the largest part of the economy was also based on slave labor performed in an agricultural setting. The result was that a master class with a particular ideology of domination, though a different one, was in charge of a slave system based on large plantations. The slaves, in turn, confronted comparable pressures and conditions of exploitation, and reacted mostly through resistance, but in some cases through rebellion.

Yet, an important difference that makes the comparison particularly enlightening is the fact that, unlike the slave societies in the Americas, the plantation system in the Sokoto Caliphate had not emerged as a direct result of capitalist market forces and did not focus specifically on the production of selected crops that commanded the capitalist world economy. Though present, the influence of the external market in the Sokoto Caliphate was minimal compared to the needs of the internal market, which absorbed virtually all the production of the main crops (grain, cotton, and indigo) produced on Sokoto plantations.[7]

In this chapter, I intend to argue that, given the existence of both striking similarities and differences, comparison between the four largest slave societies in the nineteenth-century Atlantic—the U.S. South, Cuba, Brazil, and the Sokoto Caliphate (see Map 3)—may shed further light on the nature of slave management, slaveholding ideology, and slave resistance in the "Cotton Kingdom" and its neighboring large slave economies in the Americas under the second slavery.

Map 3 The U.S. South, Cuba, and Brazil with Their Major Crop-Producing Regions Under the "Second Slavery"; and an Overview of the Sokoto Caliphate

Map 3a The U.S. South

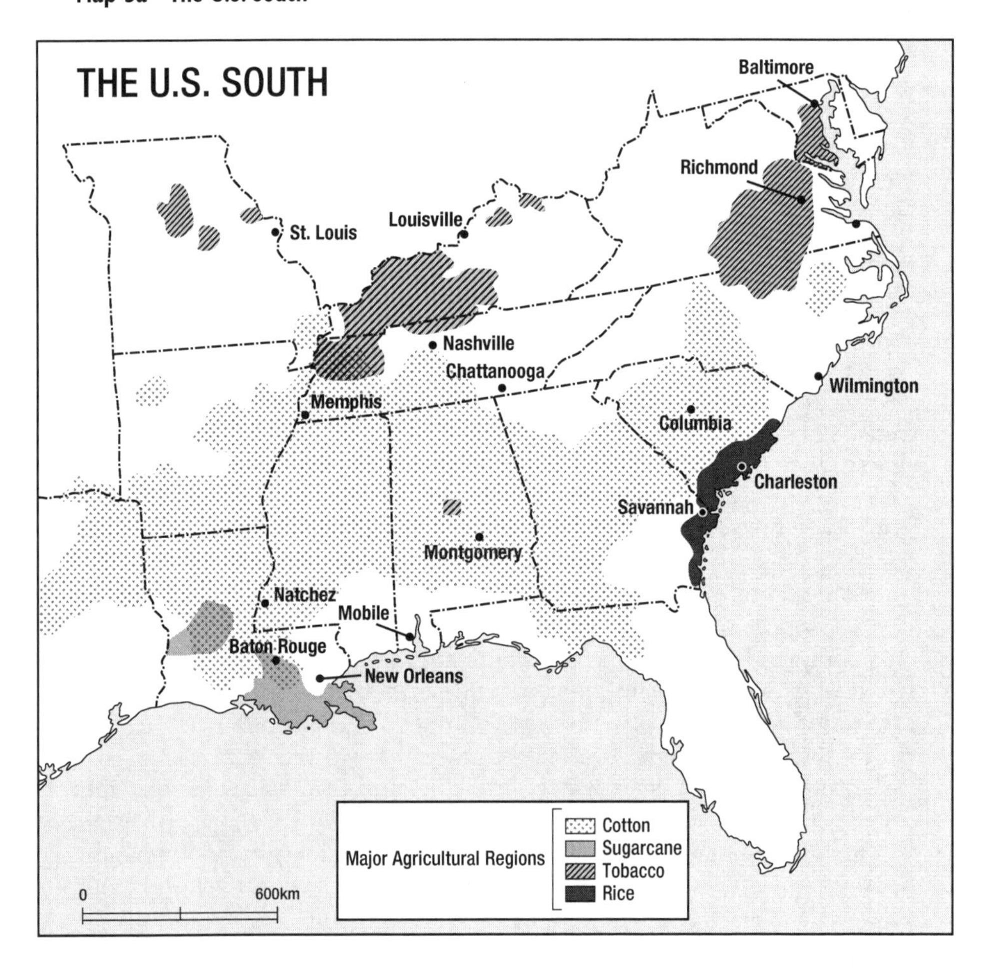

Map 3b Cuba

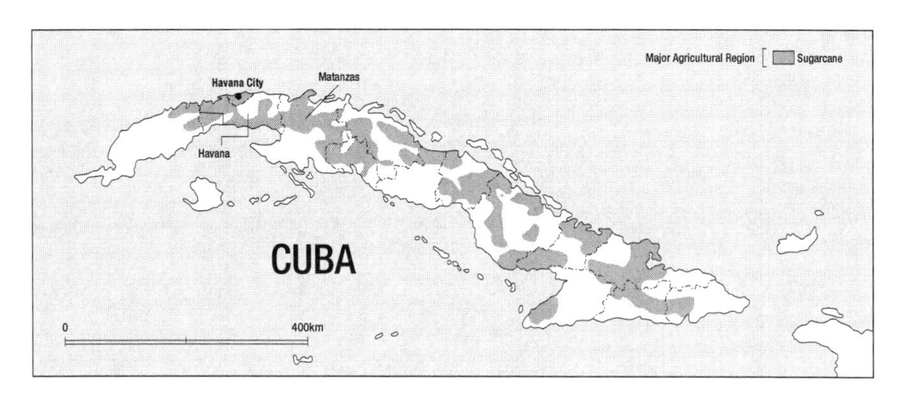

Map 3c Brazil

Map 3d The Sokoto Caliphate

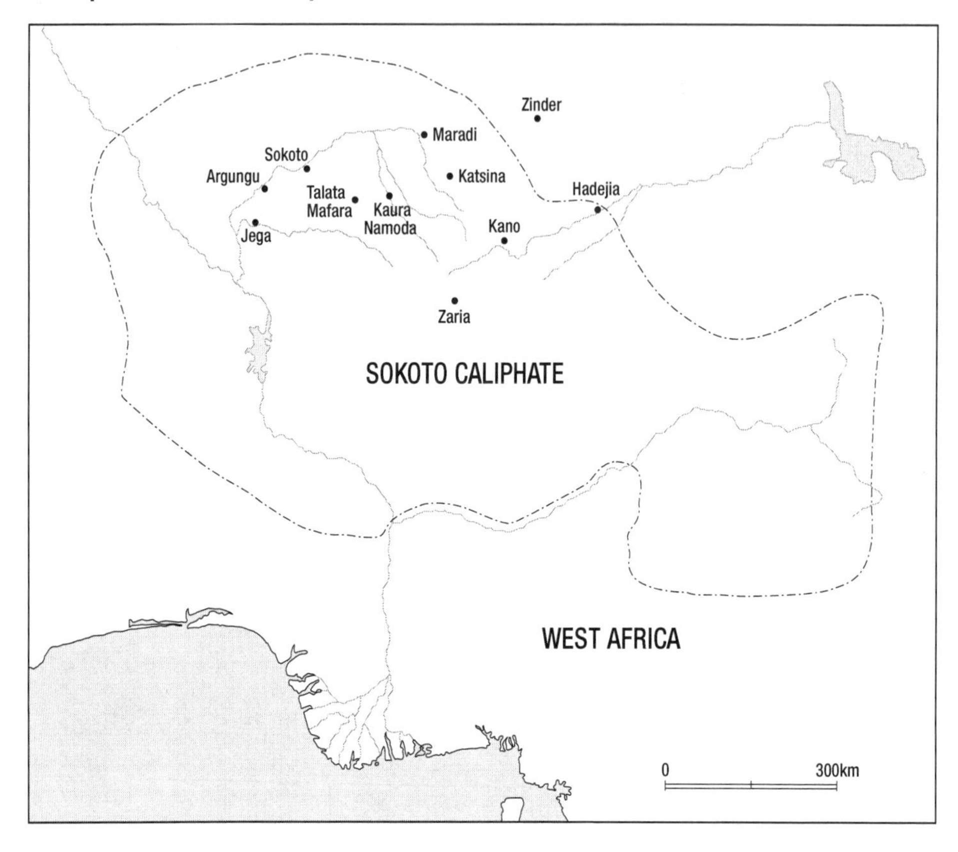

ADVANCED AND LARGE-SCALE SLAVE SOCIETIES IN THE ATLANTIC WORLD

On one hand, as part of the nineteenth-century Atlantic plantation zone and the second slavery, the U.S. South, Cuba, and Brazil shared several analogous characteristics as "advanced" slave societies. Specifically, these characteristics were the concentration on production of a single cash crop and the consequent dependence on its world-market demand, the staggering expansion in the amount of land colonized by plantation agriculture on the commodity frontier, and the planters' repeated efforts at introducing biological and technological innovation. On the other hand, it is also true that the difference in the types of crops grown in these three regions and the specific states of which they were part were largely responsible for their different histories of involvement with the world economy.

Thus, the cotton, sugar, and coffee economies occupied different niches in the nineteenth-century world economy—with different values given to the commodity frontiers that produced the different crops, according to the course of their demand in the world market. Also, the regions of production belonged, respectively, to the expanding American republic, to the declining Spanish empire, and to the rising Brazilian empire. These factors generated clear differences between these three case studies with regard to the relationship between the planters and the state, and also to the support the planters received in their projects of agricultural expansion. But there were also clear differences in terms of sheer size between the three regions, since, as Michael Zeuske reminds us, by 1850, it was clear that "the USA and Brazil were the giant slave societies, while Cuba was little more than a small part of a colonial empire."[8]

Yet, in an Atlantic perspective, next to the "giant" slave societies of the U.S. South and Brazil, and most definitely before Cuba, was the Sokoto Caliphate, the only contemporary large-scale slave society of comparable size, given the millions of slaves working in its plantations.[9]

Since the Sokoto Caliphate was not an advanced slave society with features resulting from the rise of the second slavery, it did not share several of the characteristics that the U.S. South, Brazil, and Cuba had in common—first and foremost, their focus on single cash-crop production geared toward the world market, consequent to their status as commodity frontiers in the Atlantic world. However, comparably to what happened in the three slave societies in the Americas under the second slavery, in the Sokoto Caliphate, the expansion of the slave system occurred from the beginning of the nineteenth century and accompanied a massive colonization of large regions, whose land was then incorporated within the Sokoto plantation system.

Together with the lack of focus on cash-crop production for the world market, though, another important difference characterized Sokoto agricultural expansion: the major role played by the Islamic state not only in supporting this expansion, but also specifically in granting land to free Muslim men for the purpose of establishing slave-based plantations and farms. This was a role without comparison with the slave societies in the Americas, where plantations tended to be private enterprises.[10]

The Second Slavery in the Nineteenth-Century U.S. South, Cuba, and Brazil

Of the three regions of the Americas where the second slavery flourished, the U.S. South was the one more clearly and directly linked to developments related to the Industrial Revolution. Concurrently with Eli Whitney's 1793 invention of the cotton gin, which speeded up enormously the process of production of short-staple cotton, it was the rapidly growing demand of mainly English, but also New England, textile mills that caused the enormous expansion of antebellum southern cotton production. The facts that the United States was a young republic that was protecting slavery and was determined to colonize its large territories in the West helped immensely. By the 1820s, slaveholders had expanded cotton production throughout the area between the Appalachian Mountains and the Mississippi River, while, since the beginning of the nineteenth century, they had established a flourishing sugar industry in Louisiana.[11]

In essence, what caused the U.S. planters' movement westward had a lot to do with the phenomena associated with the *market revolution*, both the early industrialization of the Northeast and the increasing linking of different areas to a national and international market through a transportation network. The market revolution was also responsible for the expanding population's pressure on the western lands, which were abundant and cheap, and could be grown afresh, unlike the burnt-out land of older eastern seaboard states. In this immense commodity frontier, both large planters and small slaveholders grew cotton in an increasingly larger number of plantations and farms, particularly alongside the navigable rivers and on the extremely fertile soil of the Deep South's Black Belt. By 1860, cotton production reached a total amount of 4.5 million bales.[12]

In comparison with the U.S. South, the history of the Cuban economy during the second slavery seems to have been similarly driven by profit-seeking planters. The planters readily used to their own advantage the structures and resources of the Spanish empire, with which they stood in an uneasy relationship. In the early nineteenth century, they sought an alliance with enlightened government officials, particularly Francisco Arango y Parreño, who effectively acted as their spokesman, with the objective of replacing St. Domingue with Cuba as the leading producer of sugar in the Caribbean. As a result, planters managed to create the legal framework and infrastructures that allowed the staggering expansion of Cuban sugar plantations, particularly in the western area of the island, south and east of the capital Havana. Here, sugar plantations (*ingenios*) quickly increased enormously in number, so that by the 1820s, Cuban production commanded, together with Jamaica, the world market of sugar. By 1850, with an output of 294,952 metric tons, Cuba alone produced almost one-quarter of the world supply of sugar.[13]

In the case of Brazil, a great deal more than in Cuba, the new era of agricultural expansion under the second slavery was tied to the logics of empire—in this case, imperial Brazil—and to rivalry with competing powers in the world market for cash crops, the most sought after of which became coffee. Prompted by the growing world demand for coffee, Brazilian planters established quickly the largest and most flourishing coffee-producing industry in the world in the three southern provinces of Minas Gerais, São Paulo, and Rio de Janeiro, with a particular dense concentration of plantations in the Paraiba Valley around Vassouras. The result was that, in the first half of the nineteenth century (and even beyond), comparably to the American cotton and Cuban sugar commodity frontiers, the Brazilian coffee commodity frontier also commanded the world market. By the 1840s, Brazil's rate of coffee production had reached such levels that its share of the world output was over 40 percent, making the South American country the leading coffee producer in the world.[14]

Plantation Organization and Agricultural Production

In the nineteenth-century Americas, and particularly in the U.S. South, Cuba, and southern Brazil, the second slavery and the expansion of plantation areas it entailed were typically associated with centralization of production activities directed from a headquarters and with organization of management through a hierarchical chain of command, even though, to a certain extent, both of these were already implemented since earlier times. In particular, in the antebellum U.S. South, on cotton plantations in the Old Southwest and sugar plantations in Louisiana, the large, often colonnaded planter mansion (the Big House, descendant of earlier structures first used on both tobacco- and rice-growing estates) was not just the dominant feature of the agricultural landscape, but also the true administrative center and point of convergence of all the economic activities.[15]

From the Big House, the planter supervised the overall functioning of agricultural production of his estate and stood at the top of a simple but effective hierarchy of management. This usually included a single overseer (given that the average number of slaves was around 40 per plantation), who effectively managed the day-to-day labor activities. Also, particularly in the cotton-producing areas, the overseer managed several slave drivers, each of whom was in charge of a gang of laborers, whose size varied according to the number of workers available. Only a few major planters who owned more than one plantation sometimes also employed a steward, who functioned as an overall administrator.[16]

Compared to the antebellum U.S. South plantation, the average Cuban *ingenio* in the region around Havana was, in general, much larger. *Ingenios* usually included at least 300 slaves, and often even more. Cuban sugar plantations shared with American plantations the presence of a distinctive administrative center, placed on the most prominent spot of the estate, where the planter (*haciendado*) lived and from where he directed agricultural production. However, the hierarchy of management was, necessarily, more complex.[17]

Manuel Moreno Fraginals called this hierarchy "a system of government." It usually included an administrator (*administrador*) who, starting from the 1820s, dealt with the plantation's economy; a general overseer (*mayoral*) in charge of slave management; and field guards or slave drivers (*contramayorales*). Also, an accountant (*mayordomo*) was in charge of food, clothing, and housing. Moreover, there were also *boyeros*, who looked after oxen, and a sugarmaster, who was in charge of the sugar mill and its workers.[18]

Unlike Cuba's estates , and similar to American plantations, Brazilian coffee plantations could be relatively small, even though still with an average of 70 to 100 slaves, since, akin to cotton and unlike sugar, coffee required little initial capital investment. On the other hand, there could also be plantations as big as the largest ones in the Paraiba Valley, near Rio de Janeiro, with 300 to 400 slaves each.[19]

In each of these plantations, usually in the most elevated spot, stood the *Casa Grande* (Big House), where the Brazilian coffee planter (*fazendeiro*) lived, and where the administrative and organizational center of the agricultural operations was based. The hierarchy of management was fairly complex, with different types of overseers (*feitores*) hired by the plantation owner. First and foremost, there was a general plantation manager (called simply *feitor*, or *administrador*). Especially in the absentee-owned plantations, the *feitor* was the one in charge of the overall economy and organization. More specific tasks related to slave management were handled by a field overseer (*feitor de partido*), a yard overseer (*feitor de terreiro*), and different types of drivers (*feitor de roca*, or *capataz*), each in charge of a gang (*terno*) of 20 to 25 slaves.[20]

Slave Management and Productivity

Clearly, the plantation systems of the Americas under the second slavery represented the emblematic examples of how, in response to market demands, pressure was put on wholly unfree laborers in terms of ever-increasing regimentation and pace of work in order to guarantee fast and efficient mass production of particular cash crops. In turn, this process led to the "modern" phenomenon of the enslaved workers' alienation, stemming from the increasing routinization of their labor. This feature shows particularly clearly in slave narratives. For example, in Solomon Northrup's *Twelve Years a Slave* (1854), the author describes very graphically the routine of labor on a sugar plantation: "three gangs are employed in the operation [of planting a sugar field]. One draws the cane from the rick, or stack ... Another gang lays the cane in the drill ... The third gang follows with hoes." Routinization of labor and the consequent alienation led to the slaves' continuous attempts to avoid the worst consequences of this condition through the various forms of resistance described in the scholarly literature.[21]

Typical was the case of the antebellum U.S. South. In all the areas of new commercial agriculture—both the Cotton Kingdom region and sugar-growing Louisiana—slaveholders implemented the exhausting gang system, as opposed to the more humane task system, in order to increase productivity on the plantations and improve rationalization of labor management. In the words of Richard Follett, "disciplined, interdependent gangs could sustain an intense and constant rhythm of work ... once equipped with an integrated order of established labor gangs, slaveholders could routinize labor while maintaining strict supervision over the slaves who toiled beneath the overseer's eye in the open field." At the same time, in their discussions over the best way of managing their plantations, slaveholders routinely advised each other on implementing the gang system, masking the system's brutality with a paternalistic rhetoric of reciprocity that overlooked slave resistance, and argued that slaveholders repaid their laborers' efforts by taking care of them and giving them food and shelter.[22]

Similarly to American slaveholders, nineteenth-century Brazilian planters in the major coffee-producing regions, such as the Paraiba Valley, in their desire to increase as much as possible the plantations' productivity, put additional pressure on their enslaved laborers through an exhausting regimentation and pace of work. However, Brazilian coffee planters adopted a mixed system with characteristics of both the gang system and the task system.[23]

On one hand, the Brazilian overseer supervised slave gangs (*ternos*) in the operations of weeding and cropping. On the other hand, the slaves' labor was also measured by the overseer individually, according to specific tasks set for each slave. The result, as Rafael Marquese has noted, was "the imposition of an incredible amount of work on slaves." Yet, we can possibly see the fact that planters paid (in monetary terms) those slaves who worked beyond the minimum tasks required from them as an attempt to maintain a fiction of reciprocity in the master-slave relationship. Also, the Brazilian slaveholding class engaged in long-term debates on the best way to manage their laborers, justifying the existence of the slave system as a paternalistic mission to civilize the latter by making them proper Brazilian subjects.[24]

Compared to the U.S. South and Brazil, nineteenth-century Cuba emerged as a slave system much less preoccupied with justifying the brutality of exploitation of agrarian laborers in the sugar plantations through increased regimentation and pace of work. Interestingly, by 1850, on Cuban sugar plantations, slaves of African descent worked side by side with Chinese *coolie* laborers, hired by Cuban planters as indentured servants in order to keep pace with the enormous expansion of sugar production.[25]

Both these categories of laborers worked in gangs (*cuadrillas*) under the strict supervision of Cuban overseers and drivers. The overseers kept discipline amidst an exhausting pace of work—a pace that had much to do with the particular requirements of a production driven by sugar mills. According to Robert Paquette, the gang's "factory-like rhythms were set by the operational capacity of mechanized mills." Discipline was kept mostly as a result of a variety of punishments that overseers and drivers were supposed to implement, according to several treatises on slave management. Strikingly, in the Cuban treatises, there was no attempt at a moral justification of slavery through a paternalistic ethos comparable to those in fashion in the U.S. South and Brazil. Perhaps this was because, as Rafael Marquese has argued, both African slaves and Chinese *coolies* mattered to Cuban masters only as a disciplined and productive workforce.[26]

Slavery in the Nineteenth-Century Sokoto Caliphate

Even though the actual process of the Sokoto Caliphate's formation in Western Sudan has little direct relation to the economic transformations causing the rise of the second slavery, the wider context of that process and its distant causes have a lot to do with changes in the mechanics of European expansion. In fact, it was the crisis of the colonial slave systems in the Americas that ultimately led to the end of the British and U.S. Atlantic slave trades in 1807 and 1808. This, in turn, produced a decrease in the international demand for West African slaves. The slaves were now available for use within the African states, making slavery a cheap and easily found source of labor and paving the way for the formation of a large-scale slave society in West Africa.[27]

Conversely, the actual process of formation of the Sokoto Caliphate, though related to these long-term causes and contemporaneous with their unfolding, has everything to do with the particular Islamic tradition that had developed in West Africa and the ongoing conflict between different ethnic groups. It was essentially a *jihad* (holy war), conducted by Fulani leader Shehu Usman dan Fodio and his successors in order to protect Muslim free men from the threat of being enslaved in regions mostly inhabited by Hausa and Yoruba people, that led to the creation of the Sokoto Caliphate, a state dominated by a mixed Fulani-Hausa aristocracy, in 1804 through 1808.[28]

Even though there was already an indigenous tradition of large-scale agriculture, plantations mostly worked by slaves became an omnipresent feature in the region of central Sudan only with the rise of the Sokoto Caliphate. As a result of the process of the formation of the Sokoto Caliphate, three different types of plantations existed: the largest ones, which belonged to the aristocracy and were tightly linked to the *jihad* and the consequent conquests; smaller plantations, which Caliphate officials had given to immigrants and commoners as prospective settlers; and official plantations, which belonged to the *emirs*—the Caliph's representatives in the Caliphate's provinces—and which only a particular category of royal slaves could administer.[29]

In general, according to Paul Lovejoy, "plantation production in the Islamic regions of the western and central Sudan was similar to that in the Americas with respect to the amount of land under cultivation and the concentration of land in units that could be worked on a large scale." The scale of slaveholding, though, was generally larger than in the antebellum U.S. South and closer to that of the large plantations of Cuba and of Brazil's Paraiba Valley. According to Sokoto-born jurist Imam Imoru, there were "slave owners in Hausaland who have purchased 100 to 200 slaves, and ... some slave owners with 400 or 500 slaves," while owners of 1,000 slaves and more were fairly unusual.[30]

Unlike most slaveholdings in the Americas under the second slavery, but similar to most landed estates in nineteenth-century Europe, plantations in the Sokoto Caliphate were organized according to a variety of different arrangements. Some of them were relatively small and self-contained; others consisted of different landholdings, often scattered, rather than all contiguous. Sokoto masters, whether they were aristocrats or commoners, could be equally resident or absentee, though the latter was more common among aristocrats. The chain of command they relied on consisted mainly of overseers (*sarkin gandu* in Hausa). These, similar to the slave drivers in the Americas, were always slaves who had earned the master's trust and could, therefore, supervise work operations on the plantation.[31]

Note that, unlike the cotton-producing regions of the antebellum U.S. South, the sugar-producing regions of Cuba and Louisiana, and the coffee-producing regions of Brazil under the second slavery, the Sokoto Caliphate did not produce any single crop that commanded the world market. The crops grown included mostly grain, cotton, tobacco, and indigo. Some of the agricultural production—as in the case of the highly profitable palm oil—found its way to long-distance trades, but most was geared toward the internal market.[32]

And yet, one of the most striking features of the Sokoto plantations was the fact that, even in the absence of the enormous pressure coming from competition in the world economy, Sokoto masters behaved in a fashion very similar to most masters in the Americas. They also organized slave labor on their plantation in gangs (*gandu* in Hausa) that worked under the overseer's watch and with an exhausting pace, even though Sokoto masters allowed slaves time for the Muslim five daily prayers. At the same time, also similar to what happened on plantations in the Americas, on Sokoto plantations, failure to work adequately led to harsh punishments for the slaves, which ranged from whipping to their actual sale.[33]

꩜

In the nineteenth-century Atlantic world, at the same time during which the three slave societies of the U.S. South, Cuba, and Brazil in the Americas expanded under the second slavery, as a result of the rise in world-market demand for valuable cash crops such as cotton, sugar, and coffee on the three commodity frontiers, another equally large-scale slave society—the Sokoto Caiphate—expanded in West Africa for entirely different reasons. Sokoto's expansion was a result of the *jihad* connected to the rise of the Caliphate, and consequently the latter emerged as a political and religious frontier rather than a commodity frontier.

Despite this fundamental difference, in all these four cases, the end result of the expanding process was the colonization of an impressive amount of new land, which was then settled according to an economic model based primarily on plantation agriculture. The large number and scale of plantations were characteristic features shared by all of the four largest slave societies of the nineteenth-century Atlantic.

However, on one hand, the tight connection with the world market and the specialization in particular cash crops led planters in the antebellum U.S. South, Cuba, and Brazil—mostly resident slaveholders—to adopt patterns of land organization that insisted on centralization of production and labor regimes that were particularly exhausting in order to maximize the output of agricultural items. On the other hand, not only the weak connection with the external market in comparison with the internal one, but also the variety of crops produced and the different origins of the plantations led Sokoto planters—many of whom were absentee aristocrats—to focus less on highly centralized models of land organization and efficiency of agricultural production than their American counterparts.

It is significant, though, that both in the large-scale slave societies of the Americas and in the Sokoto Caliphate, the exhausting gang system was the most widespread labor regime. This is a testimony that planters in most of the Atlantic world clearly considered this particularly exhausting type of labor organization the most effective way to control large numbers of slaves working in the fields.

THE MASTER CLASSES AND AGRARIAN MODERNIZATION

In the Americas, progressive slaveholders responded in comparable ways to the new challenges presented by the world market by showing an entrepreneurial attitude in their attempts to modernize agricultural production on their plantations. This process was particularly evident in the commodity frontiers of the Americas under the second slavery.

Within the process of agrarian modernization, mechanical innovation went hand in hand with biological innovation, since, as Anthony Kaye has recently argued, "while machines enabled planters to put more land into cultivation and work more slaves, manipulating the natural properties of cotton and sugar increased the crops that slaves produced per acre." At the same time, agricultural modernization also led to the implementation of factory-like devices for the marking of time, and, in general, to the flourishing of agricultural societies and of journals and pamphlets through which the master classes discussed techniques and methods of agricultural improvement.[34]

Ultimately, as Michael Zeuske has pointed out, in the nineteenth century, "slavery functioned as an agent of transfer of an economic culture" focused on agricultural modernization that characterized all the most dynamic slaveholding elites of the New World. Therefore, the spread of this specific type of economic culture clearly interested all the commodity frontiers related to the second slavery, including all the three most productive plantation economies in the nineteenth-century Americas: the U.S. South, Cuba, and Brazil.[35]

The crucial input that agrarian modernization, mechanical and biological innovation, and the adoption of techniques of scientific agriculture—especially in the antebellum U.S. South and Cuba, but also in Brazil—had on the dramatic rise in the cash-crop productivity on the new lands that planters colonized in the course of the nineteenth century stands in dramatic contrast with the basic absence of both interest and opportunities by the Sokoto planters to engage in a concerted effort at agrarian modernization on their plantations. This is all the more remarkable given the fact that, throughout the nineteenth century, the amount of land conquered and settled with plantations in the Sokoto Caliphate continued to grow in a large-scale process of agricultural expansion.[36]

And yet, in an indirect way, plantations in the Sokoto Caliphate were somewhat related to technology, particularly because, comparably to the way the U.S. South's plantations supplied the northern U.S. factories with cotton, Sokoto plantations supplied the renowned textile industry centered in the city of Kano with the indispensable indigo for the fabrication of clothes sold in long-distance trade. Therefore, despite the fact that agrarian modernization was never a feature of Sokoto plantation agriculture, it is still possible to find interesting comparative points with some features of the function of plantation agriculture in at least one of the large-scale slave societies under the second slavery in the Americas.[37]

Agrarian Modernization in the Antebellum U.S. South

In the antebellum U.S. South, technological innovation led, notably, to the widespread use of the newly invented cotton gin. Starting from the 1790s, the cotton gin quickened enormously the pace of short-staple cotton production and heralded the massive expansion of the Cotton Kingdom in the whole area of the Old Southwest, as a result of the growing demand for cotton from the textile mills located primarily in England and also in the northern United States.

Thanks to the cotton gin, starting from the original areas of cultivation in upcountry South Carolina and Georgia and around Natchez, in Mississippi, within a few decades, plantations and farms multiplied. This eventually created the largest area of short-staple cotton production in the world. Although on a smaller scale, equally important were the changes that the technological innovation brought to Louisiana's sugar cultivation. From 1822 onward, Louisiana planters' adoption of steam-powered mills increased enormously the production of sugar on each plantation.[38]

These technological innovations went hand in hand with the slaveholders' adoption of factory-like devices to enforce regimentation of labor and with improvements in the construction of infrastructures and of new means of transportation, such as steamboats, geared toward integration of the agricultural markets. At the same time, progress in biological innovation prompted slaveholders to constantly make experiments, leading to the creation of always better varieties of the premiere cash crops of cotton and sugar.[39]

As these important technological and biological innovations occurred and led to the continuous expansion of the production of cotton and sugar, the demand for laborers in the newly settled areas of the Old Southwest grew exponentially. The result was the flourishing of a massive internal slave trade—a true "second Middle Passage," in Ira Berlin's words. It sealed the fate of hundreds of thousands of African Americans, who were sold on the markets placed on the southeast-southwest route, the largest of which was New Orleans. Therefore, in the antebellum U.S. South, technological and biological innovation led at the same time to the massive expansion of plantation agriculture and to the planter class's renewed commitment to the enslavement of African Americans as the only possible means of meeting the increasing demand for labor in the Old Southwest.[40]

For those slaveholders and reformers who were at the forefront of agricultural modernization in the antebellum U.S. South, technological and biological innovation and the spread of scientific knowledge in agriculture went hand in hand with the defense of slavery. In all the leading economic journals—for example, Edmund Ruffin's *The Farmer Register* and John D. Legaré's *The Southern Agriculturalist*—the stress was equally on the improvement of farming techniques and on the efficiency of plantation management and slave supervision. Throughout the antebellum period, in every state of the U.S. South, both in specialized journals and in state agricultural societies, slaveholders reported their findings, discussed their application of scientific methods, and engaged in debates on agricultural improvement and the management of their slaves. In turn, both the journals and the agricultural societies acted as means to spread education on scientific and technological progress related to agriculture.[41]

This effort was not confined to the level of the states, since slaveholders who were actively engaged in modernizing southern agriculture effectively created an intrastate network of knowledge focused on agrarian modernization. This network found its natural locus of discussion in the major journal *De Bow's Review*. This was a publication wholly committed to the modernization of the entire southern economy, whose readership was spread throughout the U.S. South in the 1840s and 1850s.[42]

Agrarian Modernization in Nineteenth-Century Cuba

Technological innovation under the second slavery reached its peak in nineteenth-century Cuba, where, starting from the 1790s, the Cuban slave system embarked on a path on which massive expansion of sugar production and technological improvement went hand in hand. In practice, the Cuban sugar industry expanded phenomenally in a few decades as a result of the

Cuban planters' ability, in the words of Dale Tomich, "to increase the area under cultivation, establish new plantations, concentrate labor, and incorporate scientific advances into production."[43]

Thanks to their semimechanized and fully mechanized steam-powered grinding mills, by the 1840s, from their *ingenios*, Cuban planters commanded the world supply of sugar. These advancements in both technology and production occurred in conjunction with an expansion of the railroad network that connected the many sugar plantations to the island's ports. By 1860, the railroad network covered 1,281 kilometers and connected the sugar plantations in the western part of Cuba to important regional ports such as Matanzas, from which steamships also departed.[44]

Steam-powered horizontal sugar mills, as opposed to the traditional vertical *trapiches*, existed in Cuba since the 1790s. They were introduced by St. Domingue's sugar technicians who had fled the island in the wake of the Haitian Revolution, and by Arango y Parreño, who was a staunch advocate of planters' modernization and technological innovation. However, it took several decades before fully mechanized machines became more commonplace and changed, in the process, the face of Cuban sugar production. Even in the 1840s, the sugar mills on the island were, at best, semimechanized—both steam-powered and animal-powered. Two decades later, according to Reinaldo Funes Monzotes, "the sixty-four mechanized *ingenios* in 1860 were about five per cent of Cuba's sugar plantations, but their production was fifteen per cent of the total." [45]

The result of the changes brought by technological innovation led not only to the massive expansion of sugar production, but also to the deforestation and colonization of large sections of Cuba, especially in the east, for the purpose of establishing sugar plantations. It also led to the construction of major railway lines, especially in the Havana-Matanzas regions and in the Colon Plain, which became the centers of Cuba's sugar production.[46]

During these processes, Cuba's large planters, who had been mainly responsible for accelerating the pace of technological change, emerged as the privileged landowning class in a monocultural economy, in which they were the only ones who could afford the expenses brought by the adoption of mechanized sugar mills that used British-originated steam-power technology. As a result, particularly in the decades before 1860, as Dale Tomich has noted, "there was a steady process of land concentration in conjunction with the adoption of steam power and new manufacturing techniques."[47]

Agrarian Modernization in Nineteenth-Century Brazil

In contrast with Cuba, the expansion of coffee production under the second slavery in nineteenth-century southern Brazil was more the result of a shrewd adaptation of Caribbean agricultural techniques to the environment of places such as Vassouras, in the Paraiba Valley, which were particularly well suited for the establishment of large-scale coffee plantations (*fazendas*). Plantations and smaller farms were common in the other coffee-producing areas of the eastern regions of São Paulo and the western Paulista plains, and in the southeastern frontier region of Minas Gerais. In contrast with sugar, and similar to cotton, coffee cultivation did not need any expensive industrial equipment, and therefore, both planters and smaller slaveholders could begin the business and make a profit from it.[48]

Brazilian writers interested in agronomy who visited these areas in the 1830s (by which time Brazil had become the world-leading coffee producer) reported in detail about the coffee planters' skills in the choice of soils, in their preparation, and in the planting of bushes. Then, after 1850, the introduction of railroads in several areas reduced considerably the costs of transport to ports and markets, while it provided coffee planters with an increasingly integrated network of shipping and distribution. Significantly, Herbert Klein and Francisco Vidal Luna have called this crucial process of transformation of infrastructures "a revolution in transport," given that it allowed, simultaneously, the settlement of the interior areas and the massive expansion of the plantation system related to the coffee commodity frontier of the southern region of Brazil.[49]

Comparably to the sugar's parable in Cuba, coffee was also already produced in the eighteenth century in Brazil. It was the collapse of St. Domingue's economy due to the Haitian Revolution, together with the fact that Brazilian planters in Rio de Janeiro learned from both St. Domingue and Cuba about the techniques of coffee cultivation, that led to Brazil's first expansion in coffee production, prompted by Europe's growing demand.[50]

Similar to the cotton-producing regions in the U.S. South, in Brazil, biological innovation was crucial. During the first half of the nineteenth century, in the Paraiba Valley, Brazilian planters made experiments in coffee production first in mixed-crop farms, where they grew coffee plants (which took up to three years to yield beans), together with corn, and then in a monocultural regime. In time, even though not as crucial as in the case of sugar, technology also played its part. As Laird Bergad has remarked, "the application of new technologies ... in the drying and hulling of coffee was of great importance in increasing the efficiency of ... Brazilian slave production."[51]

Also comparable with the U.S. and Cuban cases was the resulting intense process of colonization of areas to establish plantations in the Paraiba Valley and in the regions of São Paulo and Minas Gerais, where coffee planters were constantly searching for the best virgin land. As a result, planters gave origin to a very flourishing internal and interregional slave trade within Brazil, akin to the U.S. South's second middle passage. Then, in the second half of the nineteenth century, the construction of railroads facilitated communication between Brazil's coffee-growing interior and the seaports with their export markets. In the process, similar to what happened in the U.S. South and Cuba, Brazilian coffee planters acquired an enormous concentration of wealth and power, which derived from their ability to grow the best quality of coffee with the best techniques of cultivation at their disposal in the best land available.[52]

Sokoto Slavery: Agricultural Expansion in the Absence of Modernization

There is little in common between the processes of large-scale technological and biological innovation that characterized the most advanced slave societies in the Americas under the second slavery and the relative absence of a tradition of scientific agriculture among planters in the Sokoto Caliphate. Clearly, in terms of both agrarian modernization and adoption of techniques of scientific agriculture, we simply cannot compare the antebellum U.S. South, Cuba, and Brazil with the Sokoto Caliphate, since all these features were crucially important in the former three, but mostly absent in the latter. The reasons for this are related, on one hand, to the different positions and roles of the three former slave societies *vis-à-vis* the latter one within the world economy, and, on the other hand, to the different origins and modes of their processes of agricultural expansion.

In the Americas under the second slavery, agricultural expansion was tightly linked to the expansion of markets on the commodity frontiers, conducted by highly commercialized and somewhat capitalist-oriented master classes. In the Sokoto Caliphate, agricultural expansion was a product of the politics of religious conquest in the name of Islam. Yet, despite this important difference, in both the cases of the Americas' slave societies and of the Sokoto Caliphate, agricultural expansion occurred on a continuous basis for the best part of the nineteenth century and on a very large scale. It led in both cases to the planters' progressive acquisition of large amounts of new land for their plantations, and, consequently, to a large increase in the number of slaves needed for agricultural work in the new areas of settlement.[53]

Starting from the period of the formation of the Sokoto Caliphate (1804 to 1808), the leaders of the *jihad* pursued a policy of expansion of the Caliphate's frontiers that kept them occupied for almost a century. In the process, they also created the preconditions for a massive expansion of plantation agriculture, even though, unlike what happened in the commodity frontiers of the Americas, this was neither the primary nor the initial concern of the master class. Expansion in the form of holy war justified the enslavement of enemies, while plantations became quickly a common feature in the new frontier areas. According to Paul Lovejoy, "under the guise of pursuing the *jihad*, slave raiding and war were institutionalized into a coercive system for the mobilization of labor." This, in turn, provided the conquerors with the enslaved workforce they needed for their newly settled plantations.[54]

Beginning with Muhammad Bello, Shehu Usman dan Fodio's son, the Sokoto Caliphs pursued a policy of encouraging the conquering Fulani population to settle in fortified centers called *ribats*. The pattern of construction of these centers had a long history behind it in Muslim lands. Their function was, effectively, to protect the plantations that formed the heart of the agricultural hinterland and the very basis of the Caliphate's economy in the frontier regions. Therefore, in the Sokoto Caliphate's case, frontier agricultural expansion in the shape of plantations continued to be strictly linked to ongoing warfare and defense needs throughout the nineteenth century.[55]

Even though there was little pursuit or interest by Sokoto planters in agrarian modernization, in one important respect a number of Sokoto plantations participated to a wider technological dimension. This was a result of the preeminence of the internal trade of goods between the Caliphate's agrarian countryside and, especially, the large city of Kano.

Kano and its southern surroundings were characterized by a flourishing textile industry. This industry involved the work of large numbers of slave women and also some men, who spun and wove large quantities of raw cotton produced on the Caliphate's plantations and transported it to the capital with very long trips. Once ready, the textile was taken to dyeing centers, and finally sold to the external market.[56]

Although a very high-skill and large-scale enterprise, the Sokoto textile industry was by no means advanced technologically, given that there were no factories or textile mills. However, from a comparative point of view, we should consider two important points. First, the fact that the textiles produced in Sokoto manufacturing centers were known and appreciated for their high quality throughout West Africa during the nineteenth century and were, effectively, the most requested Caliphate products in long-distance trade activities shows us how, indirectly, Sokoto plantations producing cotton participated to the logics of external markets wider than the regional one. Second, even though the level of technology is not comparable, it is interesting to notice that, similarly to U.S. southern plantations supplying British and northeastern American textile mills with their cotton, Sokoto plantations supplied the textile industry based in Kano (the largest manufacturing center in the entire West Africa) with cotton and indigo. In both these cases of large-scale slave societies, therefore, plantations produced raw materials indispensable for the creation of manufactured products sold on the world market.[57]

The essential premise that justifies comparison between the three preeminent nineteenth-century slave societies in the Americas and the Sokoto Caliphate is that, in absolute terms, the latter was the only comparable large-scale slave society outside the New World located in the Atlantic. Yet, in terms of the master classes' attitude toward agrarian modernization, comparison with the technologically and scientifically underdeveloped Sokoto agriculture—even taking into account the strict link between the cotton and indigo produced on Sokoto plantations and the renowned Kano textiles—highlights even more clearly that the antebellum U.S. South, Cuba, and Brazil under the second slavery were characterized by uniquely modern features. Such features were best expressed not just by the slaveholding elites' attention to technology and scientific agriculture and biological innovation, but also, especially in the antebellum U.S. South, by the fact that professional societies and journals featured extensive debates on various aspects of agrarian modernization. While a great deal of the need for modernization in plantation agriculture in the Americas had to do with the planters' competition in the world market, the modernizing features also demonstrate the existence of an essentially open attitude by many American masters toward the capitalist world economy.

UNIQUE AFRICAN AMERICAN AND AFRICAN SLAVE CULTURES

At the heart of contemporary scholarship on slaves in the slave societies under the second slavery in the Americas are two fundamental and closely related issues: the slaves' degree of retention of African cultural traits and the slaves' ability to resist the masters' attempts to exercise absolute power through acts of resistance, and eventually, organized rebellions. On both these issues, previous scholarship tended to treat the U.S. South on one side and Cuba and Brazil on the other side, as two separate and entirely different cases. In fact, in complete contrast with the situations in both Cuba and Brazil, the earlier ending of the slave trade and the much larger number of creole American-born slaves in the U.S. South seemed to have led to a high degree of absorption of the few features of African culture that were still remembered into American culture, as specifically religious practices testified. At the same time, also in complete contrast with the situations in both Cuba and Brazil, small-scale slave resistance seemed to have been the norm among the slaves, with very few large-scale conspiracies and even fewer slave rebellions on record in the U.S. South. In recent years, though, scholars have investigated more deeply both issues and have come to different conclusions.[58]

On the subject of retention of African cultural traits, several different studies focusing on Africa have demonstrated that the culture of different African groups brought to the New World, including the ones brought to North America, survived in part the African diaspora in the Atlantic, but it underwent an inevitable process of adaptation, generating new cultural forms. In her book *Slavery and African Ethnicities in the Americas* (2007), Gwendolyne Midlo Hall has argued convincingly that "specific groups of Africans made major contributions to the formation of new cultures developing throughout the Americas," including the U.S. South. Of these new and unique African American cultures, religion was clearly one of the most evident expressions.[59]

On the issue of slave resistance and rebellion, compared to Cuba and Brazil, the evidence of large-scale slave activity in the U.S. South was unimpressive. However, on one hand, it is possible to argue, as Olivier Pétré Grenouilleau has done, that day-to-day resistance and running away—both of which had comparable levels of occurrence in all the three slave societies—could have been just as effective in harming productivity and the plantation economy as a whole. On the other hand, even though only a few, the large-scale slave conspiracies and rebellions that occurred in the antebellum U.S. South show crucial points in common, particularly in regard to the role of culture and religion, with slave rebellions and conspiracies in Cuba and Brazil. These cannot be easily dismissed, and they speak of common stories of slave resistance across the Americas in need of further comparative analysis.[60]

In comparison with slave societies in the New World, at first sight the condition of slaves in the Sokoto Caliphate would seem so different as to be almost incomparable. Yet we must recognize that, even within the context of a Muslim, rather than Christian, cultural world, there were certainly comparable phenomena of slaves' assertion of their own cultural identity through the adoption of heterodox forms of religious cults. Typically, these cults were clearly distinct from the ones practiced by the masters. They included millenarian elements, pointing to possibly interesting comparisons with millenarian beliefs among slaves in the Americas.[61]

Comparison becomes even more compelling when we consider that, similarly to what happened in the Americas with cults derived in small or large parts from African traditions, in the Sokoto Caliphate, slave millenarian cults fed a series of revolts, whose influence did not stop at West Africa. In this connection, the Atlantic dimension reveals itself to be particularly important and evident. The long wave of these forms of cultural and religious resistance practiced by Sokoto slaves hit Brazil with the series of early nineteenth-century uprisings, culminating in the 1835 Malè rebellion. The protagonists of this rebellion were Muslim slaves who had arrived in Bahia from the Sokoto Caliphate through the Bight of Benin, as a result of the Atlantic slave trade.[62]

Slave Culture and Religion in the Nineteenth-Century U.S. South, Cuba, and Brazil

In investigating slave culture under the second slavery, we can begin by noticing that religion played a major role among slaves in the U.S. South, Cuba, and Brazil. In all these three areas, slave religion assumed decisively a syncretistic character through the combination of African and mostly, but not exclusively, Christian features. If this may seem to have been less evident in the antebellum U.S. South, we need to take into account the fact that, as Michael A. Gomez has argued in *Exchanging Our Country Marks* (1998), "by 1830 ... American-born slaves far outnumbered those" from Africa. And yet, while it is true that, in the decade of the 1830s "the general patterns of the emerging African American identity are discernible," this by no means signifies that African culture was not still present in some form.[63]

The 1830s were a crucial decade also because African American slaves came in contact with the revivalist type of Evangelical Christianity spread among Methodists and Baptists through the Second Great Awakening. This was a religious reform movement akin to the 1740s Great Awakening with regard to the challenge it presented to the established church authority. It focused on charismatic preaching, spontaneous mass congregation, the personal experience of God, and repentance for the purpose of salvation.[64]

As a number of studies has shown, while witnessing the phenomena associated with the Second Great Awakening, the slaves adopted many features of this particular type of Christian worship. They combined, in syncretistic rituals practiced particularly in African American Baptist churches, aspects of this specific Protestant tradition with clearly African cultural traits they had retained—specifically the singing and dancing, clapping of hands, and "call-and-response" pattern.[65]

Compared to the antebellum U.S. South, slave religion in both Cuba and Brazil showed a syncretistic character with more pronounced African elements, mainly as a consequence of the continuous imports of slaves through the Atlantic slave trade, at least until the mid-nineteenth century. The result in both cases was expressed in highly syncretistic cults that combined the worship of Catholic saints with that of African gods—particularly Yoruba gods from present-day Nigeria, given their preeminence in many areas of both slave societies—through the often open and public activities of ethnic African cultural and religious organizations; the counterpart of these simply did not exist among African American slaves in the United States. In nineteenth-century Cuba, these cultural and religious organizations were called *cabildos de naciòn*. They were officially recognized societies that functioned as points of aggregation of both free and enslaved Africans. There were, however, also secret societies, such as the Abakua, mostly of Yoruba origin.[66]

Both the *cabildos* and the secret societies practiced a syncretistic cult known as *Santería*, with a clear set of rules in terms of worshipping: the *regla de ocha*. The *ocha* were the original seven Yoruba gods, or powers, called the *orisha*, which were related to different aspects of the world, from creation to its natural constituents. These gods were assimilated to Catholic saints; hence, the name *Santería* to indicate the cult as a whole.[67]

Similar to what happened in Cuba, in nineteenth-century Brazil, as in the previous century, there were ethnic African cultural and religious societies, called *Irmandades* (Brotherhoods), in which both free and enslaved Africans, and also mulattoes, met and aggregated. The situation was somewhat more complex than in Cuba, as a result of the larger degree of variety in the population and differences in terms of class, race, and African ethnicity. The Brazilian *Irmandades* functioned mostly as types of religious brotherhoods—a long tradition of which existed already in both Portugal and Spain. The nineteenth-century Brazilian *Irmandades* practiced a highly syncretistic cult, known as *candomblé*, at the heart of which was the worship of African, not just Yoruba, gods assimilated to Catholic saints, similarly to the Cuban *Santería*.[68]

As a result of the more complex ethnic situation of Brazil, the *candomblé* religion was divided in three main strands: *candomblé ketu*, akin to *Santería*

and based on the worship of Yoruba *orixas*, as they were called in nineteenth-century Bahia; *candomblé bantu*, based on the worship of different gods of the Bantu people, a large ethnic and linguistic African group stretching from West to South Africa; and *candomblé jeje*, based on the worship of the *vodun* gods of the Fon and Ewe people of Dahomey, present-day Benin.[69]

Slave Resistance and Rebellion in the Nineteenth-Century U.S. South, Cuba, and Brazil

As several scholars have pointed out since the 1970s—first and foremost Eugene Genovese in *Roll, Jordan, Roll* (1974)—in studying the world of the slaves, religion emerges as the connecting point between culture and resistance. Religion was at once the most important form of slave cultural expression and also a highly significant way for slaves to resist the imposition of the masters' culture through the construction of one of their own. The retention of African cultural traits was a very important part of the latter process. More than this, in all the three slaves societies we have analyzed under the second slavery, religion—the unique cultural novelty constituted by syncretistic Afro-Christian cults—played a crucial part in important episodes of the most extreme types of organized resistance. This resistance was in the form of both rebellions and conspiracies, which occurred throughout the Americas at different times during the nineteenth century.[70]

Even though the social and political contexts were different in the antebellum U.S. South, Cuba, and Brazil, and the groups who initiated the revolts were also different, an undeniable similarity is the existence of an autonomous or semiautonomous tradition of slave culture and religion containing African elements. Although in different degrees, this was an indispensable precondition for the occurrence of plots and rebellions under the second slavery. The other major element in common was that many of the rebel leaders were inspired by the successful example of Haiti, whose influence in the Atlantic world the current scholarship has completely reassessed. [71]

Studies of day-to-day slave resistance in the antebellum U.S. South have a long tradition that goes back to the 1970s, when scholars realized its importance in the dynamics of the master-slave relationship. But only recently have historians reassessed the disruptive role played by runaway slaves, quantifying the number of runaway cases in the hundreds of thousands. As for organized rebellious action, though, substantially, two unsuccessful plots and two full-scale revolts stand out.[72]

In 1800, in Virginia, at a time when memories of both the Haitian Revolution and the American Revolution were still vivid, a slave named Gabriel planned to march with an army of slaves, and possibly also poor whites, into the capital Richmond and force the white elite to live up to the ideals of freedom and equality contained in the Declaration of Independence. However, he was discovered and executed.[73]

Also probably inspired by the Haitian Revolution was the revolt that took place in Louisiana in 1811. The revolt involved at least 200 slaves, and led to several killings and the burning of a few plantations before it was stopped.[74]

Ten years later, in Charleston, South Carolina, a free colored man named Denmark Vesey planned to unite both free and enslaved African Americans in a rebellious act that, on the night of July 14, 1822, would have allowed the conspirators to flee to the safe haven of Haiti. This plot was also discovered, and the trial records show, undeniably, what a crucial role syncretistic African American religion played in it. These records repeatedly refer to Vesey's second in command, an individual who came from the Gullah—a slave group that, given its particular condition of isolation in South Carolina's Sea Islands, retained many African religious and cultural traits. In practice, as Douglas Egerton has remarked, "Vesey and his chief lieutenant, 'Gullah' Jack Pritchard, an East African priest, fused African theology with the Old Testament God of wrath and justice."[75]

Religion played an even more crucial role in the events that occurred in Southampton County, Virginia, in 1831. At that time, a slave named Nat Turner succeeded in raising a small army of slaves and killed 55 whites before being found, tried, and hanged together with 17 of his 70 followers. Born with remarkable gifts and convinced of being special and chosen by God, Nat Turner managed to convince his followers that his mission was one infused with millennial elements: to accelerate the end of times, and bring about the end of slavery and the final judgment for white Virginians. These were themes that, undoubtedly, Turner refashioned syncretistically from revivalist meetings related to the Second Great Awakening.[76]

Interestingly, after 1831, there were no more notable slave revolts in the antebellum U.S. South. An important coincidence is that, according to Michael Gomez, in the 1830s, African cultural and religious traits began to lose their strength in the United States, as a result of the increasingly larger number of creole American-born slaves over African-born slaves. Even though this is a matter for speculation, perhaps this loss of cultural strength also had an impact on the ability of these traits to inspire slaves to stage rebellions.[77]

In comparison with the antebellum U.S. South, in nineteenth-century Cuba slave resistance was not just commonplace, but also led to the formation of temporary maroon communities in remote areas of the countryside, which were called *palenques* and were inhabited by *cimarrones* (runaway slaves). Also in Cuba at that time, the Haitian Revolution acted as a constant inspiration for rebel slaves, while the slaves' strong African cultural and religious identity also acted as an important factor in the wave of conspiracies and rebellions that swept over the island between 1802 and 1844.[78]

In 1812, Cuban colonial authorities discovered a major conspiracy, named after José Antonio Aponte, a free colored Havana inhabitant. Syncretistic Afro-Caribbean religions clearly played an important role, since Aponte led one of the powerful Yoruba *cabildos de naciòn*, which functioned as ethnic African groups mixing free blacks and slaves, and also as religious associations connected to *Santería*. Despite the fact that they succeeded in destroying two sugar plantations, Aponte and his followers were prevented from acting further, and were captured and hanged.[79]

A number of slave rebellions occurred between the 1820s and the 1830s in the western part of the island, around Havana and Matanzas. These were all seemingly connected with the dramatic rise in the import of African slaves, mostly former subjects of the Oyo Empire, in present-day Nigeria. Between 1843 and 1844, thousands of African-born slaves rose repeatedly in Matanzas and its surroundings, burning sugar mills in a series of revolts that climaxed with the 1844 La Escalera conspiracy. However, Cuba's colonial authorities treated all these events as if they were part of a single plot—one masterminded in order to join in rebellious activities free and enslaved blacks, mulattoes, and even British abolitionists. The authorities proceeded to implement harsh preemptive measures on a massive scale. Also as a consequence of these measures, Cuba did not witness major slave rebellions again until the 1860s.[80]

Due to the larger spaces and the greater opportunities for escaping in the wilderness, in Brazil, the runaway slave communities, called *quilombos*, or *mocambos*, lasted much longer than in Cuba. Therefore, they played a major role in the assertion of the slaves' African identity and in supporting slave resistance and rebellion. Also, in nineteenth-century Brazil, as in both the U.S. South and Cuba, the Haitian Revolution was a powerful influence and inspiration on slave rebellions, at least between the end of the eighteenth century and the beginning of the nineteenth century, as Bahia's 1798 Tailors' Conspiracy testifies.[81]

In the early decades of the nineteenth century, the new rise of Brazilian sugar production, due to St. Domingue's collapse, and the changing politics of the African states led to a large increase in the imports of Yoruba and Bantu slaves from the Bight of Benin for the sugar-producing region of Bahia. There, together with the Hausa, they formed the largest groups of African-born slaves. In Bahia, especially in the capital Salvador and in the Reconçavo, the longest and largest wave of Brazilian slave conspiracies and rebellions occurred between 1807 and 1835. In all of them, African-born slaves were protagonists against free blacks and mulattoes, while both *quilombos* and syncretistic Afro-Brazilian cults also played major roles.[82]

In the early conspiracies and revolts between 1807 and 1816, the year of a major uprising with plantations burned and whites murdered, Hausa slaves were the protagonists. In the final and largest uprising in Bahia in 1835, however, the leaders of the rebellion were all Yoruba Muslim slaves, called *Malé*. And yet, even though original Malé Muslim practices were distinctively African, in Salvador, they mingled syncretistically with *candomblé* practices. Non-Muslims also took part in the rebellion scheduled for January 25, 1835. According to Laird Bergad, the rebellion was "designed to secure freedom for the slaves of

Bahia, especially those of African birth." However, the Brazilian authorities, alerted to the possibility of revolts, forced the conspirators into premature action, defeating them in battle, and subsequently unleashed a massive wave of repression. Similarly to the Cuban case, these measures prevented the occurrence of major slave rebellions in Brazil in the following years.[83]

Slave Religion, Resistance, and Rebellion in the Nineteenth-Century Sokoto Caliphate

Comparably to what happened within the slave communities of the antebellum U.S. South, Cuba, and Brazil under the second slavery, in the Sokoto Caliphate, religion played a major role in the assertion of the slaves' identity, and therefore, ultimately served as a form of cultural resistance to the masters' authority. It was all the more so in the Sokoto Caliphate because of the centrality of Muslim religion in West Africa and Central Sudan and also because, from the times of Sokoto's early formation onward, the Caliphs had pursued a policy of expansion in the guise of *jihad* against non-Muslims.[84]

In the early years of the Caliphate's expansion, slaves who proclaimed their allegiance to the *jihad* were given freedom, but, as the Caliphate settled, after the conquest of the central Hausa emirates, slaves could have a prospect of freedom only if they declared their allegiance to the Caliphate's religious orthodoxy, which many of them were not prepared to do. Hence, over the course of the nineteenth century, an increasing number of Sokoto slaves joined heterodox forms of Islamic religion, and particularly millennial types of reform movements, clearly as a form of resistance not unlike the one experienced though syncretistic Afro-Christian cults by slaves in the Americas.[85]

Millennial movements in West Africa were, typically, forms of Mahdism—the belief in the coming of the Mahdi, the restorer of Islamic faith and justice. By far the most important of these movements was associated with the Isawa sect, which, according to Paul Lovejoy, "like Mahdism, promised a purge of the world, only in the name of Isa (Jesus)." Thus, we could reasonably say that the Isawa sect, which was popular among the Sokoto slaves since it appealed particularly to them, represented a form of religious syncretism that combined elements of Islam and Christianity.[86]

In turn, a millennial cult like the Isawa sect produced important leaders such as Hamza in the 1840s and Malam Ibrahim in the 1850s, both in the region of Kano, who led thousands of Sokoto slaves in large-scale slave revolts in opposition to the Caliphate. It is remarkable that all the episodes of large-scale slave revolts that plagued the Sokoto Caliphate during the century of its existence were connected to Islamic millennial movements, and therefore had a strong element in the tight link between the slaves' assertion of a religious identity independent from the one that characterized the master class and the slaves' quest for freedom. To a certain extent, this important link characterized also the Sokoto slaves who ended up in Bahia. Therefore, an understanding of the West African context of religious resistance can help us to better comprehend the occurrence of the early nineteenth-century slave revolts culminating in the 1835 Malê uprising.[87]

In fact, the thousands of Yoruba slaves who arrived in Bahia in the 1820s and 1830s had been made captives in the course of the *jihad* that had led to the expansion and consolidation of the Sokoto Caliphate. This occurred particularly in the southern areas, close to the Bight of Benin, from where Sokoto slave traders—among the most active in the entire West Africa—had sent their newly captured slaves to replenish the Atlantic slave trade and the plantations of the New World. Therefore, the Yoruba slaves, some of whom were Muslims, had been opponents to the *jihad* in Africa. However, once they found themselves in Brazil, the only strong common element of identity they could use to oppose their new masters became effectively their shared experience of the *jihad*. This gave rise to the creation of a common Islamic, though also highly syncretistic, affiliation among the Yoruba slaves in Bahia, and remarkably, the 1835 Malê uprising followed a similar pattern to the *jihad* that characterized Sokoto expansion, almost to the point of being a transatlantic extension of the latter in the Americas, as some scholars have claimed.[88]

The Americas Under the Second Slavery and the Sokoto Caliphate in West Africa

A broader comparative perspective on the antebellum U.S. South within the wider context of the nineteenth-century Atlantic world can help us to gain insights into a number of issues related to perennial questions on American slavery. First and foremost is the question of its modern features in relative and absolute terms. In this respect, looking at the antebellum U.S. South in comparison with Cuba and Brazil, the other two large-scale slave societies in the Americas interested by the phenomenon of the second slavery, helps us not only to understand better the deep ties of nineteenth-century plantation agriculture in the Cotton Kingdom and in the analogous sugar and coffee commodity frontiers with the capitalist world economy, but also to highlight the particularities of the specific modern features that characterized all three case studies. These include features such as the masters' open attitude toward agrarian modernization, with particular attention to technological and biological innovation, and the adoption of scientific techniques of agriculture, together with the flourishing of professional agricultural societies and journals.

Strikingly, comparison with the only other effectively comparable large-scale slave society in the Atlantic world outside the American continent, the Sokoto Caliphate, confirms beyond a doubt the modernity of the New World planters' general economic attitude, particularly in the antebellum U.S. South. This shows especially in their efforts to improve plantation agriculture, mainly as a result of the much tighter relationship of the slave societies in the Americas with the world market, when compared to the less market-oriented and more religious and political conditions of planters in West Africa. Yet, it is also important to remember that, despite the enormous differences, planters of all four large-scale slave societies in the nineteenth-century Atlantic seemed to have privileged a similar system of slave labor organized in gangs, perhaps for both similar and different reasons.

In turn, slaves working on the plantations of all four large-scale slave societies in the Atlantic world resisted slavery in a variety of ways, all of them geared toward the assertion of the slaves' identity in face of the masters' repeated attempts to annihilate it. What makes resistance among slaves in the antebellum U.S. South, Cuba, and Brazil unique, though, is the fact that slaves in all these societies asserted their own identity by creating particular cultural forms of religious worship. They formed novel syncretistic cults, which combined African and Christian elements in different ways and degrees according to the different regions where they lived.

By the nineteenth century, whether the memory of African practices was only vague, as in the antebellum U.S. South, or still very strong, as in Brazil and Cuba, slave identity was based a great deal on the assertion of novel cultural forms in which either hidden or evident African cultural memories were present. In turn, this process of assertion of unique cultural identities also fed resistance all the way to the more extreme forms of rebellion, as the importance of the religious element in a number of episodes of slave revolts in the three major slave societies in the nineteenth-century Americas attests.

Even though it might seem hardly the case, it is possible to make a comparison with analogous phenomena of slave resistance and rebellion in the Sokoto Caliphate. There, slaves also asserted their particular identity through forms of worship different from the ones of their masters, though within the context of a Muslim rather than Christian culture, as the famous episode of Sokoto Muslim slaves revolting in the Malé rebellion in Bahia in 1835 attests. That event shows particularly well the importance of understanding the Atlantic context when studying comparative slavery in the Americas.

Part B

AMERICAN SLAVERY IN THE EURO-AMERICAN WORLD

Chapter Four

SERVITUDE AND AGRARIAN LABOR IN THE EURO-AMERICAN WORLD

In the first half of the nineteenth century, agricultural labor in wide areas of the American hemisphere and the eastern and southern European peripheries was mostly associated with large landed estates. In all these areas, production of specific crops for sale in the world market led to the implementation of rigidly regimented systems of labor with different degrees of *unfreedom*—from slavery to serfdom, to various forms of sharecropping and tenancy. In other words, during the course of the nineteenth century, all these areas were commodity frontiers in the sense of the expression used by Jason Moore within a framework referring to Immanuel Wallerstein's model of analysis of the historical expansion of the European world-economy.[1]

Following Moore's suggestions, by applying his model to the nineteenth century Euro-American world, in this chapter, we will look at the commodity frontiers of the Americas—in particular, the antebellum U.S. South—in comparative perspective with the European commodity frontiers of grain, olive oil, citrus, and wine grapes. My aim is to suggest possible comparative points focusing on similarities and differences in the organization of production and in work management between the two systems of large landed properties—American plantations and European landed estates—which were responsible, in different ways and degrees, for the expansion of capitalist space in the agricultural areas of the two continents.

As a result of concurrent processes of expansion of the capitalist space in the commodity frontiers of the Americas and Europe, the rise of the *second slavery* in the Atlantic world occurred in parallel with the renewal of the *second serfdom.* This expression relates to the original process of consolidation of most of the eastern European *latifundia*—large landed estates—which had taken place between the sixteenth and the seventeenth centuries. In the first half of the nineteenth century, in the territories stretching from Prussia to Russia, large-scale landed estates produced grain for the world market through the employment of a harsh type of unfree labor characterized by either a serf or a semi-serf legal status. At the same time, grain was also one of the main staple crops, along with olives and citrus, that characterized agricultural production in the *latifundia* and landed estates of southern Italy and Spain. Here, even though Italian and Spanish peasants were technically free, labor management took the form of a variety of exploitative sharecropping arrangements.[2]

Altogether, the large landed properties and the *latifundia* present in the eastern and southern European regions went through a process of reconversion comparable to that of Atlantic plantations during the second slavery. This led to a renewed emphasis on production for the world market at the beginning of the nineteenth century.

From the 1980s onward, starting with Peter Kolchin's seminal work, scholars have engaged in sustained comparison of the antebellum American South, as a specific area in the Atlantic world, with both eastern and southern European regions in the nineteenth century, focusing on slavery, serfdom, sharecropping, and tenancy. They also have studied the ideology of the landed elites at the time of the renewal of the second serfdom in the decades of 1800 to the 1860s, which was the period between the beginning of the nineteenth century and the world-significant events leading to the emancipation of American slaves and Russian serfs.[3]

In this chapter, I will provide some suggestions on themes for further studies by relying mainly on the work of the few comparative historians who have engaged in this specific type of comparison. By focusing on Russia, Prussia, southern Italy, and Spain—four European regions in which nineteenth-century agricultural transformations were particularly incisive—I will explore possible avenues of comparison specifically in the areas of land management, labor control, and agrarian modernization.

I will not confine my analysis to strictly comparative studies, and I will make wide use of the most established and up-to-date scholarship available on nineteenth-century agrarian labor, agricultural economy and society, and elite ideology in the four regions I mentioned. However, I will still rely a great deal on the three studies that have dealt specifically with American slavery and eastern European serfdom, and with American slavery and southern European sharecropping and tenancy in comparative perspective: Peter Kolchin's *Unfree Labor: American Slavery and Russian Serfdom* (1987), Shearer Davis Bowman's *Masters and Lords: Mid-Nineteenth-Century U.S. Planters and Prussian Junkers* (1993), and Enrico Dal Lago's *Agrarian Elites:*

American Slaveholders and Southern Italian Landowners, 1815–1861 (2005). No study has yet been published on a comparison between American slavery and nineteenth-century Spanish agrarian systems of labor.[4]

The aim of this chapter is to suggest that we should see slavery in the antebellum American South not just as part of the Atlantic world's plantation system that underwent the changes associated with the second slavery, but also as part of a much wider context of changes that occurred in the commodity frontiers of the world economy on a global scale as a result of the influence of the Industrial Revolution. These changes affected in comparable terms two different peripheral regions characterized by the predominance of agrarian economies: the Americas (north and south) and eastern and southern Europe (see Map 4). In both these regions, the demand for raw materials coming mainly from industrialized England and also, on a lesser scale, from a few industrializing countries, led to comparable transformations in the large landed estates' emphasis on production of those crops—old and new—that commanded the nineteenth-century world market. In turn, this move toward commercialization of agriculture geared to production for the world market affected, in comparable ways, the conditions of agrarian laborers, while it prompted a similar drive toward modernization of agricultural production.

Map 4 Major Crop-Producing Regions in the Americas and in Eastern and Southern Europe (1800–1860)

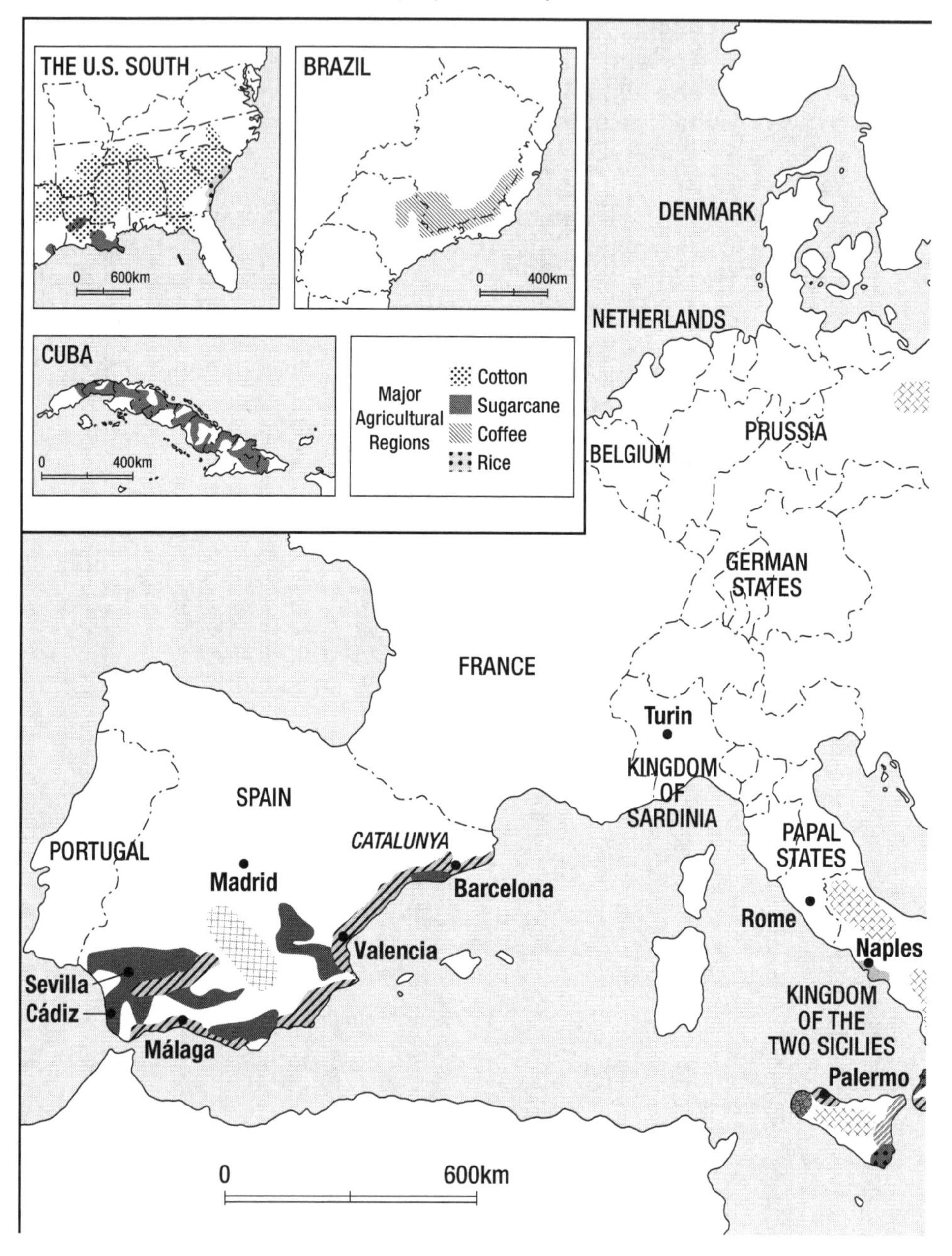

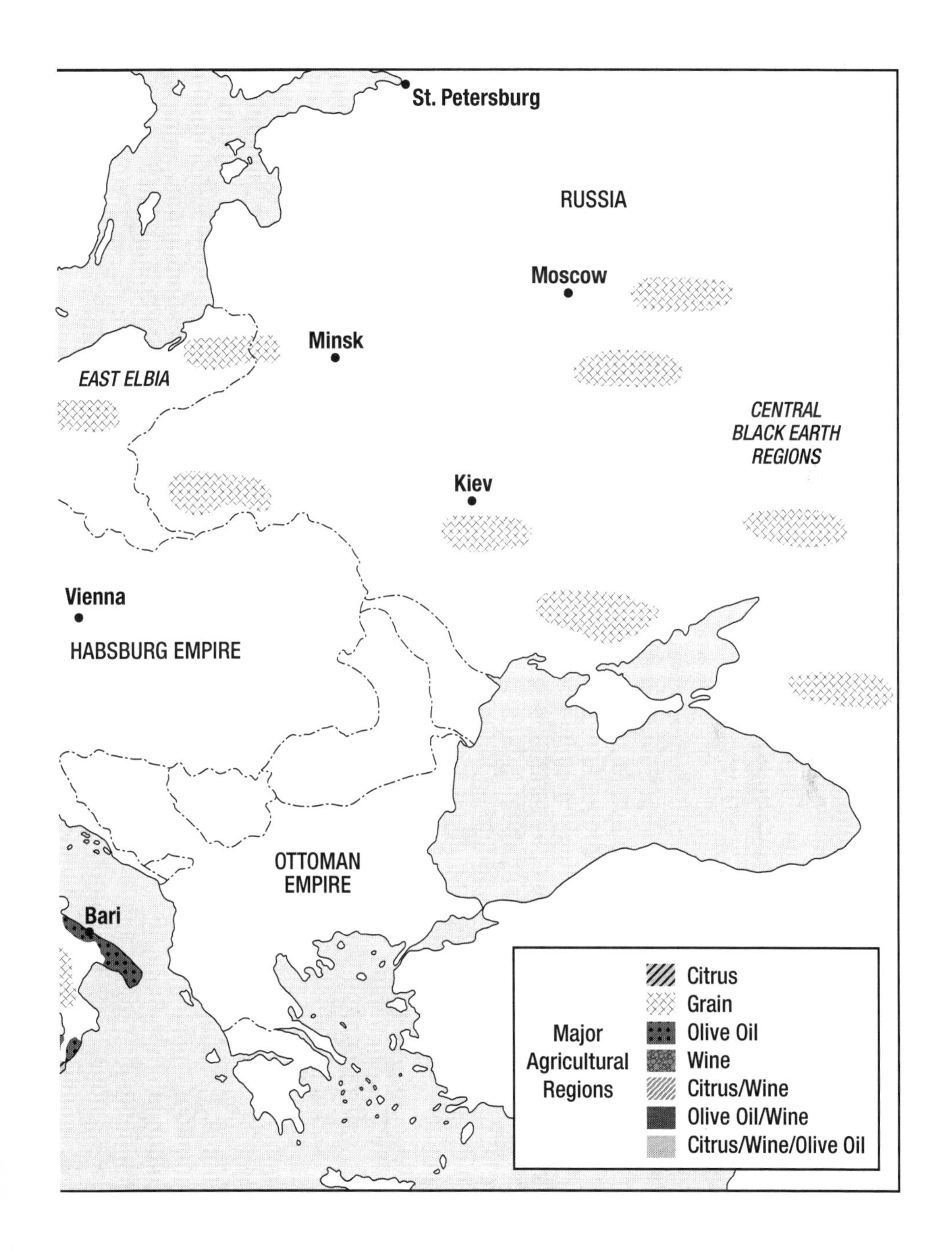
St. Petersburg
RUSSIA
Moscow
Minsk
EAST ELBIA
CENTRAL
BLACK EARTH
REGIONS
Kiev
Vienna
HABSBURG EMPIRE
OTTOMAN
EMPIRE
Bari
Major
Agricultural
Regions
Citrus
Grain
Olive Oil
Wine
Citrus/Wine
Olive Oil/Wine
Citrus/Wine/Olive Oil

EURO-AMERICAN SYSTEMS OF SERVITUDE AND AGRARIAN LABOR

Historians use the collective term *servitude* to indicate a broad category that includes all systems of unfree labor, and therefore both American slavery and European serfdom. According to Michael L. Bush, servitude is, first and foremost, "a legal institution," and as a consequence, the name indicates a legal condition of "lack of freedom, in the sense that the servile were bound to obey to whoever owned their time and labor."[5]

Bush, whose *Servitude in Modern Times* (2000), remains the most comprehensive general study on modern systems of unfree labor, recognizes that "servitude came in a number of quite different forms, some of them far more exploitative than others." This is, effectively, the point of departure for recognizing that American slavery and European serfdom were different systems of unfree labor, with different degrees and modes of exploitation, which, nonetheless, were but varieties, or types, of a general category of servitude. Therefore, they are eminently comparable.[6]

Here, I wish to go further still and argue that different types of servitude, or different types of unfree labor, can be fruitfully compared with forms of legally, or nominally, free labor. I argue this specifically in relation to the agrarian world, where most of the types of servitude originated. In this world, even when technically free, ex-slaves and ex-peasants continued to be bound to both landowners and their land, in different degrees and ways, by contractual obligations and informal requirements. They effectively ended up being free only to a limited extent. This was certainly the case of tenants and sharecroppers in many areas of Mediterranean Europe, and particularly in southern Italy and Spain. Therefore, this is the rationale for comparing American slavery with southern European sharecropping and tenancy, however different and distant from one another the two systems might seem at first sight.

Euro-American Servitude in the Transition to the Nineteenth Century

In order to start our comparative analysis of American slavery with eastern European serfdom and southern European sharecropping, we need to first look back at the early modern period. This was roughly the same time when plantations in the Atlantic underwent their first phase of colonial slavery, while also landed estates in eastern and southern Europe underwent a radical transformation. In eastern Europe, this transformation was related to the rise of the so-called *second serfdom.*

As Robert Brenner has pointed out, it was "the problem of the low and declining labor-land ratio," together with "the opportunity to profit from the export of grain to growing markets in western Europe" that prompted eastern European landlords to tie their peasants to the land, and specifically to the production of grain in large landed estates, at a time (the sixteenth century) when serfdom had died out in most of the western part of the continent. *Latifundia* (large landed estates) also characterized southern Europe, where serfdom survived only as a legal definition, and peasants were mostly free and landless laborers. Here also, however, the growing market for grain produced important changes in the early modern period. Among them was an increasing commercialization of agricultural production. Effectively, we can consider the entire area of production of grain in large landed estates, stretching from Russia to southern Italy and southern Spain, as an area akin to the plantation zone in the Americas between the seventeenth and the eighteenth centuries.[7]

Comparatively speaking, by the 1820s, the Americas and Europe appeared transformed in almost every respect economically and socially. The political upheavals brought by the Age of Revolutions had led to the destruction of a number of old agrarian regimes, while the mechanisms of the world market had led to the rise of new ones. In the Atlantic slave system, a process that had started with the abolition of the slave trade, first by Britain (1807) and then by the recently formed United States (1808), reached its peak with the emancipation of slaves in the British colonies in 1833 and in the French colonies in 1848. In Europe, the Napoleonic regime had abolished feudalism in most of the western part of the continent. In 1848, the legal termination of serfdom reached Prussia and the Habsburg lands.

Seen from a comparative perspective, as in Michael Bush's *Servitude in Modern Times,* the Euro-American world at mid-century appeared as a place where freedom was rapidly advancing and was increasingly cornering unfree agrarian labor from every direction. Nonetheless, those places where slave societies continued to flourish in the Americas—the U.S. South, Cuba, and Brazil—and also several places where serf societies disappeared at a later time—notably, central Russia—were hardly on the verge of social and economic collapse. Also, the abolition of feudalism and serfdom in areas such as southern Europe, much like the abolition of slavery in the British colonies, had led to the employment of systems of only nominally free wage labor, in which emancipated agrarian workers were exploited in some cases to a greater extent than before.[8]

The nineteenth century saw the peak of the Industrial Revolution, led by Britain as the hegemonic world economic power. The demand for new types of raw materials prompted by the Industrial Revolution had comparable effects in the Americas and Europe, and, naturally, on New World slave and postslave societies and on European serf and post-feudal societies. Such effects included both a movement toward centralization of production around large landed estates and an increase in the level of economic exploitation of agrarian workers, free and unfree, for the purpose of boosting the production of particular items.

Thus, the nineteenth century saw the rise of the second slavery and the golden age of cotton production for most of the U.S. South, of sugar for Cuba and Louisiana, and of coffee for Brazil. Comparably, serf- and non-serf-based production of grain reached very high levels in several eastern European areas, including Prussia and Russia, while production of commercial crops did the same in southern European regions such as Italy's south, or *Mezzogiorno*, and coastal Spain. While some general works have looked at the global context in which these changes have occurred, a few more specific and explicit comparisons between case studies situated in the New World and Old World have done much to clarify how these same changes affected areas as diverse as elite ideology, work management, and workers' lives.[9]

Agricultural Peripheries and World-Market Demand

It is significant that Dale Tomich and Michael Zeuske relate their concept of second slavery to that of second serfdom as a way to point out a process of "systemic redeployment" and "expansion." If that was true for the plantation zone in the Americas, it was all the more so for the landed estate areas in Europe. In other words, concurrently to the changes related to the rise of the second slavery between the end of the eighteenth century and the beginning of the nineteenth century, crucial transformations, mainly brought by the indirect influence of industrialization on the peripheral and nonindustrialized areas of Europe, occurred also in the regions characterized by the second serfdom and by production conducted on *latifundia*.[10]

In practice, the changes that the restructuring of the European large landed estates entailed were related to a global reconfiguration of the world economy. From the 1780s onward, similar to the Americas' plantation regions under the second slavery, Europe's peripheral areas functioned mainly as suppliers of agricultural products for England (the country the Industrial Revolution had made the "workshop of the world") and for a few other areas, among them the northern United States. Both the industrialized and the industrializing countries' demand for raw materials generated an expansion of production of both old crops—in the case of Prussia and Russia—and new crops—in the case of southern Italy and coastal Spain—in the commodity frontiers of peripheral eastern and southern Europe. This led to a reconfiguration of the agricultural economy, which in all four cases relied on large landed estates to meet effectively the rising world-market demand.[11]

In eastern Europe, particularly important developments in regard to changes in the second serfdom occurred in Russia. There, a combination of factors, especially the industrializing western Europe's ever-rising need for wheat and the continuous expansion of land under the control of the Russian Empire, led to a massive increase in the production of grain, with a special focus on the Central Agricultural regions and the Black Earth provinces. Russia's consequent long period of command of the world market of grain lasted until the later part of the nineteenth century. As a result, in the first half of the nineteenth century, commercial farming became increasingly widespread in the heartland of the grain-producing region, in the central and southern steppes. This brought a change in the perception of serfs, now considered more a source of labor than of income, and the rise in the amount and intensity of *corvée* labor demanded from them.[12]

Compared to what happened in Russia and eastern Europe, the phenomenon of adjustment to the globalization of an agricultural market commanded by England and the industrializing countries led to completely different, but no less important, types of changes in southern Europe. In particular, starting from the end of the eighteenth century, in southern Italy and southern Spain, several large areas that continued to be characterized by the presence of *latifundia* mainly grown with grain, and also other areas, began commercial production of specific cash crops. In general terms, often, the more commercialized type of agriculture was associated with and practiced by a rising middle class of landed proprietors who had benefited from the legal abolition of feudalism in southern Italy and Spain at the beginning of the nineteenth century.[13]

In southern Italy, the new type of commercial agriculture focused on production of olive oil, citrus, and wine. Production occurred in lands at the margins of the *latifundia* and in other types of large landed estates in both the continental Italian South and Sicily. It rose constantly in the period leading to the mid-nineteenth century, and it typically employed free peasants, either as sharecroppers or tenants. Similarly, in Spain, a new type of commercial agriculture appeared particularly with the rise of wine making in Catalonia and near Cadiz and of orange growing in Valencia, and also with the expansion of olive cultivation in eastern Andalusia. All these productions grew steadily during the course of the nineteenth century. In a comparable way to southern Italy, their growth affected the choice of workers' arrangements, from sharecropping to tenancy, on the large landed estates.[14]

As a result of the restructuring of the world economy between the late eighteenth century and early nineteenth century and of the world market's need for different types of crops, the plantations in the Atlantic world and the large landed estates in Europe assumed structural features that made them better suited for efficient and organized production. In both cases, centralization of production activities was directed from a headquarters. Organization of management through a hierarchical chain of command, even though already implemented to a certain extent, became commonplace in all the regions that had a particularly tight relationship with the world market.

The differences between the agricultural systems were equally important, and had a great deal to do with the different origins of American plantations and European landed estates. On one hand, as Edgar Thompson reminds us, originally the European landed estate produced "to consume and [was] therefore organized around a variety of agricultural and animal goods yielded up to the lord as rent." Therefore, even though, in the nineteenth century, several European landed estates produced crops for the world market, commercial agricultural operations and the centralized and regimented type of land and work management that came with them inserted themselves within an older and less market-oriented structure. Conversely, from the outset, the American "plantation [was] organized [specifically] around the production of a staple crop" for the world market. As a consequence, the centralized and regimented type of land and work management associated with the nineteenth-century agrarian economy inserted itself within a tradition of commercial agriculture that, from the beginning, was not just a natural component, but the fundamental feature of plantations throughout the Americas.[15]

Thus, the nineteenth-century world market's demand for old and new crops grown in the Atlantic and European agrarian peripheries had a decisively transformational influence on the reorganization of plantations and landed estates, with a renewed emphasis on centralization of production and hierarchy of management. At the same time, it also had a profound impact on the reshaping of labor relations and labor management, both unfree and free. In both the plantation zone in the Atlantic world and the large landed estate area in Europe, the world economy's need for efficient mass production of particular agricultural items led slaveholders and landowners in the commercial agricultural areas to put additional pressure on their agrarian laborers in order to gain the maximum profit from their plantations and landed estates. The nature and degree of this additional pressure differed markedly according mainly to two variables: the particular type of cash crop produced on the estate and the degree of nominal/legal freedom laborers were allowed in a particular socioeconomic system.[16]

Agrarian Modernization and Agronomy

The restructuring of the nineteenth-century world economy, with the need for efficient mass production of particular agricultural items, old and new, had its most far-reaching consequences in the movement toward agrarian modernization that invested the entire Euro-American world and led to the rise in the importance of scientific agriculture. In the plantation zone in the Atlantic world and in the large landed estate area in Europe, where commercial agriculture was present, so was a drive toward agrarian modernization through technological innovation and scientific methods of cultivation. In practice, progressive slaveholders and landowners in the two peripheral regions of the world economy responded in comparable ways to the new challenges presented by the world market by showing an entrepreneurial attitude in their attempts to modernize agricultural production on their plantations and landed estates.

Comparably to the planter elites in the plantation zone in the Atlantic world under the second slavery, the serf-owning and landowning elites of the large landed estate area of eastern and southern Europe engaged in agrarian modernization in response to the nineteenth-century world-market demand for particular items in an attempt to step up production through the implementation of scientific agriculture and, to a lesser degree, of technological innovation. This attitude was common to most of the progressive landowners in the European peripheries. It focused particularly on the importance of agronomy as a lively object of debate in agricultural societies and agricultural publications.

Originating in northwestern Europe—specifically England, France, and Germany—between the end of the eighteenth century and the beginning of the nineteenth century, a discipline of agronomy became quickly linked to soil science and scientific techniques of improvement in cultivation. This arose in close relation with the political economy of agriculture and with rational management of the land. It was particularly spread through the writings of such luminaries as Arthur Young, Albrecht Daniel Thaer, Johan Heinrich Von Tunen, and Jean Antoine Chaptal. As a result, in the first half of the nineteenth century, the names of these particular agronomists were familiar to all those European landowners "from Ireland to Russia," who, according to Marta Petrusewicz, "competed to increase their output and worked to adopt new products and new technologies, to develop rural industries, to improve animal breeds, to introduce new crop rotations." This was especially the case in those areas, such as Central Russia and the Mediterranean coasts of southern Italy and Spain, where entrepreneurial property holders engaged in commercial agriculture and horticulture.[17]

Whether or not we wish to consider the agronomic cultures that characterized the most active landowning elites in Europe and the most active and successful slaveholding elites in the Americas as a sign of modernity, the undeniable truth is that the ultimate reason for agricultural improvement in many commodity frontiers in the peripheral areas of the nineteenth-century world economy had much to do with a chain of changes that had spread through the world market as a whole as a result of the Industrial Revolution. The effects of these changes showed particularly clearly in the area of work management, both unfree and free.

Being fully aware of the enormous differences between them, it is very difficult to compare explicitly slave management in the nineteenth-century Americas with post-feudal or even serf management in nineteenth-century Europe, let alone with industrial work management. And yet, comparison can still offer us important insights, especially on the Industrial Revolution's effects on peripheral agrarian areas of the world economy. For example, consider how Mark Smith has been able to demonstrate that the U.S. planters' imposition of strict time schedules over their enslaved workforce was part of a general attempt by the capitalist master classes, including British industrialists, to instill a "clock consciousness" in their workers, whether these workers were free or unfree.[18]

This particular comparison could easily extend, with the proper recognition of important differences, to the coffee and sugar plantations areas of Brazil and Cuba in the Americas, and also to several particularly productive agricultural areas of southern and eastern Europe. In fact, those specific comparative studies between the nineteenth-century U.S. South—often taken as a paradigm, according to Peter Kolchin, for New World slave systems, and even for slavery as a whole—and eastern Europe and southern Italy have widened the contours of the debate over the capitalist versus noncapitalist character of American planters. These studies have tended to take a position in the debate by claiming the existence of similarities, amidst significant differences, in the mind and behavior of American planters and either paternalist Russian masters or capitalist Prussian *Junkers* (landed aristocrats), or both paternalist and capitalist southern Italian landowners.[19]

Therefore, it is important to keep in mind while reading the few explicit comparative studies that deal with American slavery and either eastern European serfdom or southern European sharecropping that the scholars who wrote them wished to contribute to resolving the two conundrums related to the nature of the labor system and to the ideology of the master class in the U.S. South. They did this by implementing a novel comparative perspective, which might have allowed them to discover previously overlooked similarities and differences between New World and Old World systems of legal and nominal servitude, or between societies characterized by unfree labor, such as the U.S. South, and societies characterized by either unfree labor of a different type or by nominally free labor.

AMERICAN SLAVERY AND EASTERN EUROPEAN SERFDOM

Within the broad category of servitude, or unfree labor, slavery and serfdom had some basic characteristics in common. Chief among them was the fact that, in Michael Bush's words, workers "shared a servile condition that was not only for life but also passed to the descendants." Also, as we have seen, specifically in North America and eastern Europe, slavery and the second serfdom began as systems of unfree labor roughly at the same time in the early modern period, and as a result of similar concurrent circumstances: an abundance of land to be grown, combined with a shortage of workforce to grow it.[20]

Beyond these comparable points, scholars are divided in regard to the similarities and differences between American slavery and eastern European serfdom. The two most important comparative studies focusing on the antebellum U.S. South and Russia and Prussia—Peter Kolchin's *Unfree Labor* and Shearer Davis Bowman's *Masters and Lords*—have offered diametrically opposite interpretations of unfree labor systems, of master-bondsmen relationships, and of elite ideologies.

Following Genovese, Kolchin has argued that broad similarities existed between the antebellum U.S. South and nineteenth-century Russia in the fundamental contradiction between the commercial orientation, with the obvious differences of production of cotton and sugar on one side and grain on the other side, for the world market, and the pre-capitalist and paternalist character of labor relations. Kolchin has elaborated further on this latter point, arguing that different types of paternalism characterized the ideologies of the master classes and the labor relations in the two regions. According to Kolchin, paternalism was stronger in the antebellum U.S. South, where most slaveholders had few slaves and constant contact with their enslaved workforce, and weaker in nineteenth-century Russia, where noblemen owned thousands of serfs, whom they rarely saw, and lived mostly as absentee proprietors. This difference explained the stronger autonomy of the Russian peasant community compared with the American slave community.[21]

Conversely, focusing on comparison between planters in the antebellum U.S. South and *Junkers* in Prussia, and specifically East Elbia, in the years before 1848 and the definitive abolition of serfdom, Bowman has argued that the two elites shared not only an attention to the commercial orientation of their agricultural enterprises and the market value of their crops—cotton/sugar and grain—but also clear capitalist features in their ideologies and labor relations. Bowman clearly has followed at least some of the premises of Robert Fogel and Stanley Engerman's and James Oakes's views of American plantations as "factories in the fields." Therefore, he has treated antebellum U.S. planters and nineteenth-century Prussian *Junkers* more as full-fledged modern types of capitalist entrepreneurs, managing their enslaved and serf workforces as if in a firm, than as large landowners and labor owners struggling to do their best to exploit the commercial possibilities of the crops they grew from their peripheral positions in the world economy.[22]

Land Management and Administration in Nineteenth-Century Russia and Prussia

Given that Kolchin's and Bowman's views of the similarities and differences between American slavery and the Russian and Prussian varieties of serfdom have led them to diametrically opposite conclusions, what is the best way to proceed in order to provide a balanced view of comparison of slavery in North America with serfdom in eastern Europe? I believe we should focus on three basic themes: administration, labor relations, and elite ideology. I will suggest possible comparative points for each of them in turn, in reference to both Russia and Prussia.

First, we should recall briefly a few important features of administration and labor relations in the antebellum U.S. South under the second slavery. In regard to administration, we have noticed the existence of widespread models of structural organization of the landscapes of plantations for the efficient production of particular agricultural items, with the presence of a clear administrative center in the Big House, and the reliance of slaveholders on hierarchical chains of command, based on the triad of planter-overseer-drivers. In regard to labor relations, we have seen how the development of management practices in the U.S. South, Cuba, and Brazil led specifically to additional pressure on slaves in the form of an increase in regimentation and pace of work. This came through the employment of the exhausting gang system of labor, which was used as a result of the need to increase production and work efficiency in relation to those cash crops that met the world-market demand.

Next, it is important to acknowledge, in a comparison between American slavery and eastern European serfdom, that, unlike the lands with plantations in the U.S. South and the Americas under the second slavery, many of the new lands touched by agricultural expansion in southern and eastern Europe in the nineteenth century usually were parts of very large and diverse landed properties, several of which were scattered and not even neighboring one another. This was the result of a centuries-old policy of accumulation of land by the propertied nobility, which, especially in the eastern and southern European regions, was a tiny class in possession of most of the lands in the country in which it resided. As a consequence, unlike most planters in the Americas, most large landed proprietors in eastern and southern Europe tended to own more than one estate, to be absentee owners, and to rely on a more complex bureaucracy. However, it is also true that, in both eastern and southern Europe, from the late eighteenth century onward, resident landowning began to characterize a minority of progressive landed proprietors, who were particularly interested in exploiting the new market opportunities by actively engaging in commercial agriculture.[23]

In nineteenth-century Russia, noble landowners owned thousands of serfs and several estates. Each of these estates had its own administrative center for coordination of the business of agricultural production, which the landowners used only in those cases in which they were resident rather than absentee. This occurred mostly in the grain-growing Black Earth and Central Agricultural regions characterized by land expansion and commercial farming. Typically, the Russian noble landowners (*pomeshchiki*) relied on seigneurial stewards, whose official names varied according to the estates and the areas, together with a figure of sheriff/magistrate (*ispravnik*), especially in case of unrest. Also involved were a series of other intermediate figures who came from the local peasant community, the most important of whom were the community's head (*starosta*) and the office clerk (*kontorshchik*). Differently from other countries where serfdom was present, Russia, uniquely, had peasant communities with a large degree of autonomy, despite the peasants' serf status.[24]

Unlike what happened in Russia, in early nineteenth-century Prussia, *Junkers* no longer owned their serfs since the passage of the laws on serf emancipation of 1807 through 1810. However, the Prussian landed aristocracy continued to hold large landed estates (*Rittergüter*)—though most not as large as the Russian ones—worked by peasants bound to the *Junkers* by labor obligations. In particular, *Junkers* in East Elbia ran their *Rittergüter* as commercial enterprises, which produced grain for the world market. They mostly either directly managed or directly participated in the management of their estates. The fact that August Meitzen characterized the *Rittergut* as "a state within a state"—an expression that Shearer Davis Bowman has compared to William Smith's description of antebellum U.S. slavery as "*imperium in imperio,* a government within a government"—shows how much the concept of self-sufficient, autocratic, and well-ordered management was associated with the *Junkers*' estates. And this concept, in turn, relied on a highly centralized administration, which the *Junker* headed, and which had its second most important element in the overseer (*Inspektor*), who was the actual supervisor of the workers' performance in the fields in the *Rittergut.*[25]

Work Management in Russia and Prussia

To a certain extent, we could apply a similar reasoning in regard to laborers in the regions of the Atlantic plantation zone in the Americas and also in the eastern European areas with large landed estates where commercial agriculture was flourishing in the nineteenth century. In both cases, in principle at least, the laborers mattered to the landowners in terms of the labor that they performed on the estates. However, despite the fact that, comparably to enslaved African American workers in the antebellum U.S. South, eastern European peasants were still serfs legally bound to the land, radically different modes of exploitation characterized labor relations in eastern Europe. The cause of their implementation was often the world market's demand for particular agricultural products. In many ways, Russia was an emblematic case in eastern Europe, since exploitation of serfs effectively occurred in two ways that were common in all the areas where the second serfdom was still present, and partly even where only a few remnants of serfdom still existed in the first half of the nineteenth century, as in Prussia.

In the nineteenth century, Russian serfs needed to pay their owner money obligations (*obrok*) in arbitrarily set amounts, or else they had to perform *corvèe* labor (*barshchina*) three to five days a week. Comparably to how American slaves felt about the task system as opposed to the gang system, Russian serfs preferred the *obrok* over the *barshchina,* because the latter required them to perform labor under the strict supervision of the owner, his administrator, or his overseer. Indeed, there is a long-standing debate on the supposed enlightened—we could say paternalistic—character of some late eighteenth-century noblemen who systematically implemented *obrok* on their estates. This way, they left more autonomy to their peasants and allowed them to build a sort of serf economy, comparable to the slave economies present in task system areas of the Americas.[26]

However, note that, in the nineteenth century, in those areas where production for the market was commonplace, as in the grain-growing Central Agricultural and Black Earth regions, the *pomeshchiki* strove, as Peter Kolchin has noted, "to take advantage of it [the market production] by putting their serfs on *barshchina* and maximizing seigneurial output." This tendency increased steadily in those regions during the nineteenth century. If, on one hand, as some scholars have argued, the implementation of *barshchina* led to a more likely resident landownership, on the other hand, the effect on the serf force was an additional amount of pressure on peasants, with close supervision of their labor and strict discipline kept through customary punishments.[27]

Unlike Russia, early nineteenth-century East Elbia had witnessed, through the emancipation laws of 1807 through 1810, the effective release of the agrarian workforce from most of its legal obligations. However, the employment of *corvées* continued until 1848. Moreover, even though the same laws had also freed the land from its feudal status, making it available for purchase, a great deal of it fell in the hands of the *Junkers,* along with the land of a large number of Prussian peasant families whose farms were too small to support them. This left the peasants with no option but to sell the land and work as landless laborers for the *Junkers.*[28]

According to Bowman, from the late eighteenth century, "improving *Junkers* [those who engaged in modernization of production] came to rely less on compulsory labor services (*Fronarbeit*) provided by peasant households ... instead they made greater use of hired labor (*Lohnarbeit*) rendered by landless men and women ... subordinate to the noble landowner's manorial authority." By the first half of the nineteenth century, landless day laborers living in cottages on the *Junkers'* estates, called "cottagers" (*Isten*), were the most common type of workforce in East Elbia. They worked under strict supervision from sunup to sundown for six days a week. In return, they were given the cottage, some cash, part of the harvest from the fields, and part of the grain to be threshed in winter, effectively as if they were almost sharecroppers.[29]

The story of East Elbian peasants in the first half of the nineteenth century is a tale of the rise from serfdom to progressive emancipation and wage labor, and of adaptation to the consequences of the *Junkers'* reaction in terms of changing patterns of labor relationship. This led to a process of negotiation between the noble landowner and their *Inspektor* on one side, and their peasant workforce on the other side, over the meaning and extent of freedom.[30]

Effectively, the Russian and Prussian case studies in terms of estate administration and relations between elites and workers enlighten us on two different stages of development of eastern Europe's second serfdom. In Russia, the serf system still existed legally and bound the peasants strictly to both land and landowner. In Prussia, the system had been legally abolished, together with the entire feudal apparatus. In comparison with Russia, the first half of the nineteenth century in Prussia was a situation of transition from serf to wage labor, with the landless peasants resisting the landowners' attempts to keep full authority over them. It is the nature of this struggle, which we will also see in the southern Italian and Spanish case studies, that makes comparison with the master-serf relationship in Russia and with the master-slave relationship in the antebellum U.S. South worthwhile.

In fact, it is possible to think in all these cases—keeping in mind the different contexts and the different degrees of workers' freedom—about masters and laborers as if they were involved in processes of continuous negotiation. These processes had a great deal to do with the particular needs of specific agricultural operations and with the well functioning of the agrarian enterprises based on U.S. plantations, Russian properties, and Prussian *Rittergüter.* Clearly, the outcome of this negotiation, in the shape of a common ground between masters and laborers, was especially crucial in those areas where agrarian production on plantations and large landed estates was particularly geared to the world market. These areas included the cotton- and sugar-producing regions of the antebellum U.S. South, the Black Earth regions in Russia, and the East Elbia provinces in Prussia.

Landowners and Agrarian Modernization in Russia and Prussia

In this connection, it is particularly important to investigate in comparative perspective to what extent the antebellum U.S. South's planter elite, the Russian serf-owning elite, and the Prussian East Elbia's landowners contributed to the profitability of their regions by introducing improvements aimed at the modernization of their landed estates. Typically, these improvements related to a widespread nineteenth-century culture of modernization. In similar ways among the agrarian elites who controlled the land-based economy of the peripheries of the Euro-American world, this culture expressed itself essentially through a renewed emphasis on some basic concepts aimed at ensuring a better performance of the landed estates in the world market. Such concepts contemplated, in particular, the idea of rational management and administration of land and workers, with the important corollary of the need for the landowners' residence; the implementation of agronomy and scientific techniques of cultivation and farming; the circulation of ideas through participation to specialized agricultural societies and the publication of agricultural journals; and the creation of a professional body of either farmers or overseers and administrators through schools built specifically for their training. Either a few or most of these concepts and features were undoubtedly issues of discussion and led to specific actions taken by the most active sections of the landowning elites of peripheral regions of the Euro-American world, and therefore also of the antebellum U.S. South, Russia, and Prussia.[31]

As we have seen, several studies focusing on the ideology of the planter elites in the most profitable slave societies of the nineteenth-century Americas have shown that, from the early 1800s onward, slaveholders in the U.S. South, Cuba, and Brazil sought different ways to improve the productivity of their plantations. Through cultural transfers, the New World planter elites learned of particular techniques already successfully tested in other plantation areas of the Americas or on landed estates in Europe. At the same time, a plethora of treatises, pamphlets, journals, and other publications—sponsored or published by regional or national agrarian institutions, written by well-known progressive members of the elite, or written by agricultural reformers—played a very important part in the making of an innovative and forward-looking attitude among the American slaveholding elite. This is similar to what happened in peripheral areas of eastern Europe with a particularly tight relationship with the nineteenth-century world market, such as central Russia and Prussia.

For example, in nineteenth-century Russia, the best-known progressive noblemen who engaged in scientific agriculture were the so-called Slavophiles, who were the members of a movement that idealized Russian peasant life. Yet, starting from the 1840s, they studied the most advanced techniques of farming in the leading European countries and sought to import them to Russia, thus establishing, in the process, the first seeds of a Russian agronomist tradition. Among the leading Russian Slavophiles were Aleksey Khoyakov, who introduced a variant of crop rotation together with different types of fertilizers, and Aleksandr Koshelev, who, in the 1850s, brought to Russian agriculture technological innovation in the form of various types of machines. Similar to progressive landowners in other areas of Europe, Russian Slavophiles often owned model farms in which they implemented agricultural modernization and scientific cultivation on a relatively large scale, while they engaged in debates on the rational management of their landed estates, particularly in the occasions of scientific meetings.[32]

Compared to Russia, Prussia, as part of the larger German cultural world with all its connections with north-central Europe, had a much longer and stronger tradition in agronomic studies. The simple fact that one of the foremost nineteenth-century European agronomists was Prussian Albrecht Thaer, author of the widely read work *Grundsätze der rationellen Landwirthschaft* (Principles of Rational Agriculture, 1809–1812) shows how much agronomy was rooted in Prussia.[33]

Prussian *Junkers* were very much part of this regional agronomic culture, and several of them—such as Johann Goettlieb Koppe, Carl von Wuffen-Pietzpuhl, and the famous Otto von Bismarck—either inspired or taught by Thaer, engaged in all sorts of experiments on their landed properties. They also wrote important treatises on scientific agriculture and rational management as ways to improve crop production on Prussian *Rittergüter.* And similarly to other modernizing landed elites, *Junkers* interested in agronomy debated in the various regional agricultural associations—such as the Pomeranian Economic Society, in which Bismarck became involved—and in prestigious agricultural journals, the most famous of which was the *Allgemeine Agrarwissenschaften* (Annals of Agriculture), published in Berlin from 1815 onward.[34]

To summarize, in comparable ways to the case of American slaveholders, eastern European landowners' treatises and pamphlets focused on the implementation of scientific agriculture and on the different ways new agronomic techniques could help to improve production. This was particularly the case in relation to those specific crops that made the fortunes of the commodity frontiers and of their progressive elites in the peripheral regions of the world economy.

American slaveholders participated with other slaveholding elites in the New World in a common continent-wide movement toward the scientific and technical improvement of production on the plantations. Similarly, the Russian and Prussian landowning elites, in different ways and degrees and with a different pace, participated in a common European agronomic culture. The main tenets of European scientific cultivation and rational management related to the content of treatises written by prominent nineteenth-century agronomists, such as German Albrecht Thaer, whose name was well known among all modern European landowners.

AMERICAN SLAVERY AND MEDITERRANEAN SHARECROPPING

There is little doubt that comparison between the type of slavery practiced in the antebellum U.S. South and the type of labor arrangements, particularly sharecropping and tenancy, practiced in southern Europe—specifically, in southern Italy and Spain—adds a new dimension to comparative historical studies of slavery in the Americas.

Unlike Russia, where serfdom dominated agrarian labor relations until 1861, and Prussia, which was in a situation of transition from serf to wage labor up until 1848, Italy's *Mezzogiorno* and Spain had seen serfdom and the peasants' legal obligations connected with the feudal system abolished since the early 1800s. As a result, nineteenth-century southern Italian and Spanish peasants were technically, or nominally, free.[35]

Even more than in the case of Russia and Prussia, therefore, comparison between the antebellum U.S. South and southern Italy and Spain must take into account the fundamental differences that arise from the study of slave versus nonslave societies and labor systems. In this respect, through the careful analysis of both similarities and differences, comparison can shed light on issues such as the types of administrative systems employed in American plantations and southern Italian and Spanish *latifundia*; the varieties of management employed to control enslaved and free laborers; and the American slaveholding and the southern Italian and Spanish landowning elites' efforts at implementing modernization in their regions' agriculture.

Comparison is particularly fruitful if we keep in mind Tomich and Zeuske's framework. According to this framework, changes associated with the rise of the second slavery in the Americas were part of a phenomenon of global transformation of world capitalism. This affected all the peripheral regions by transforming them into new types of Moore's commodity frontiers that produced particular items for the world market, and specifically for industrialized England and the industrializing northeastern United States. Therefore, the basis for comparison between American slavery and southern Italian and Spanish sharecropping and tenancy in the nineteenth century is the fact that, similar to the antebellum U.S. South, Central Russia, and Prussia, the Italian *Mezzogiorno* and Spain functioned as commodity frontiers that produced particular agricultural items for the world market. They produced grain, but above all, olive oil, citrus, and wine—the three items associated with commercial agriculture in the Mediterranean.[36]

Land Management and Administration in Nineteenth-Century Southern Italy and Spain

In both southern Italy and Spain, different sections within the elites were responsible for the production and sale of the commercial crops. This was different from the case of the slaveholding elite that was responsible for the production of cotton and sugar in the antebellum U.S. South. In fact, despite the fact that it was somewhat divided along class lines, especially between the smaller slaveholders and yeomen and the larger planters, the slaveholding class in the antebellum U.S. South was remarkably homogenous by southern European standards. Even the changes that occurred with the rise of a new and increasingly larger body of planters with the expansion of cotton and sugar production under the second slavery had not seriously compromised this general homogeneity in class terms.

By the same token, the type of administrative system found throughout the plantations in the areas of cotton and sugar production, with the Big House as the operational headquarters and a simple hierarchy formed by the triad planter-overseer-drivers, was a constant feature. This system was strictly related to the need for the efficiency of agricultural performance in an extremely valuable commodity frontier, as the antebellum U.S. South was.

In comparison with the slaveholding elite of the antebellum U.S. South, the landowning elite of the Italian *Mezzogiorno* was a much more composite class. In southern Italy, more fluid social conditions than the ones in Russia and Prussia, resulting from the early nineteenth-century abolition of the feudal system, led to a sharp distinction between the absentee noble landowners (*aristocrazia terriera*) and the resident noble and newly established bourgeois landowners (generically called *possidenti*). The latter, though still a minority, were mostly present in the areas of commercial agriculture on the coasts of Campania, Apulia, Calabria, and Sicily. Here, resident landowners often used the *masserie*—the equivalent of the Big Houses in the southern Italian *latifundia*—as true administrative structures from which they directed production of olive oil, citrus, and wine, mostly through different types of lease contracts with the peasants. In general, southern Italian landowners relied on stewards (*massari*), or, if they owned many estates, on administrators (*amministratori*)and, especially in Sicily, on rent collectors (*gabellotti*), who tended to be the ones who dealt directly with tenants and sharecroppers.[37]

In Spain, similar social conditions to those in Italy led to a distinction between absentee and resident landowners that cut across class distinctions between the old and new nobility and the newly established bourgeoisie, even though in somewhat different ways. This was a result of a more gradual process of abolition of the feudal system, which was completed only by the 1840s.[38]

In general, the Spanish large landowners (*latifundistas*) tended to be absentee and rely on administrators (*aperadores* or *gerentes*), who took care of vast amounts of lands scattered in different areas, as in the case of Andalusia, and leased them to tenants (*arrendadores*). However, the situation was different when the landowners came from members of the bourgeoisie who previously had been tenants or farmers (*labradorers*) and had moved up the social ladder to become agrarian entrepreneurs in the areas of commercial agriculture, particularly in wine-producing Catalonia and Cadiz and in orange-growing Valencia. Here, small landholding, together with lease contracts, was more diffused. In these cases, landowners tended to be resident proprietors actively engaging in market transactions.[39]

Work Management in Southern Italy and Spain

The world-market demand, which was responsible for the flourishing of commercial agriculture on the Mediterranean coasts of southern Italy and Spain, was also the force that drove the cotton and sugar economies in the antebellum U.S. South. Additionally, it was also the main factor that influenced the changes in the shape and pace of management of the enslaved workforce on the plantations, as happened in all the commodity frontiers of the Americas under the second slavery. The result was that in all the regions of cotton and sugar production in the antebellum U.S. South, the exhausting gang system was the most widespread mode of labor.

However, we should remember that, as we have seen earlier, the nineteenth-century U.S. planter class's common belief in the ideology of paternalism, combined with constant resistance manifested by the slaves against the masters' pretensions of absolute power, had led to forms of contractual relationship between masters and slaves. Even though very different in shape and context, the forms of contractual relationship arising in a situation of slave labor are worth comparing with forms of contractual relationship between masters and workers that arose in a situation of legally free labor, such as the ones of the agrarian worlds of nineteenth-century southern Italy and Spain.

Unlike either American slaves or Russian serfs, southern Italian and Spanish peasants in the nineteenth century were legally free. Therefore, the nature and degree of their exploitation in terms of the agricultural labor required from them by the landowners differed radically in more than one way.

In nineteenth-century southern Italy, the chronic absence of landownership among peasants had created a very large reservoir of a workforce formed by landless laborers (*braccianti*), who often ended up either as casual workers or as contract tenants on the large grain-growing *latifundia*. In the areas characterized by commercial agriculture, or horticulture—the orange groves, the olive tree fields, and the vineyards on the coasts of both the continental *Mezzogiorno* and Sicily—sharecropping tenants rented plots from the landowners through improvement leases (*contratti a miglioria*), which could easily become instruments of exploitation in the hands of ruthless proprietors. In practice, the tenants kept the land for 10 to 15 years; after that, just as the plants began to have fruits, they had to return the land to the landowners. As Augusto Placanica noted, this is a perfect example of how "the need for specialized cultivation pushed the landed proprietors to place the burden of innovation on the laborers." Still, while that need was a result of the new world-market demand for those particular crops, in the hands of more enlightened or paternalistic landowners, the improvement leases could also become examples of reciprocity between masters and laborers. For example, this is the case of the more equitable 20-year contracts present in most citrus groves, hailed by contemporaries as examples of progressive agrarian relations.[40]

Similar to southern Italy, agriculture in Spain also suffered from a chronic absence of landownership among the peasant class. Effectively, there were a large number of landless day laborers (*jornaleros* or *eventuales*), who were employed on the grain-growing *latifundia* according to the need at particular times of the year, such as in relation to harvest. However, the situation was different in those areas of Spanish commercial agriculture that were tightly linked to the world market, specifically the regions of wine production in Catalonia and near Cadiz and of orange production near Valencia, where tenants enjoyed more opportunities in their relationship with the landowners who owned particularly profitable estates.[41]

In particular, nineteenth-century Catalonia was strikingly advanced in this sense. In its vineyards, tenants were employed with a particular type of contract (*rabassa morta*), according to which the sharecropper (*rabasser*) planted the vines and legally owned them for an indefinite duration. Moreover, tenants could keep up to two-thirds of the harvest. Not surprisingly hailed as a model agrarian contract and a true example of reciprocal relations between landowners and tenants, the *rabassa morta* lasted until the late nineteenth-century further expansion of Catalan viticulture, when the consequent aggrandizement of landed properties made it obsolete. However, until then, this type of contract provided an interesting example of response to market demand—in this case, focused on wine production—according to which landowners and peasants agreed on a less exploitative type of solution to the problem of maximizing profits coming from commercial agriculture.[42]

Landowners and Agrarian Modernization in Southern Italy and Spain

An important issue that has engaged scholars for a number of years, and that a comparative study between the antebellum U.S. South and Mediterranean regions characterized by commercial agriculture can help to solve, regards whether the attitude of American slaveholders toward improvement of agricultural production can be seen as truly modern. This is of particular interest when the American slaveholders' attitude is analyzed in comparative perspective with the attitudes of southern Italian or Spanish landowners, who dealt with a technically free labor force. The crux of this matter is the fact that the American slaveholders' modern stand *vis-à-vis* improvement in plantation agriculture led to those same slaveholders' reactionary defense of the superiority of the slave system as a whole over systems of free labor. Note that this modern stand, for Eugene Genovese's followers, did not affect the pre-bourgeois and paternalistic character of the master-slave relationship, while, for neoclassical economic historians, it was part of a general capitalistic ethos.

In this sense, the comparative perspective with the post-feudal agrarian regions of Europe can help by alerting us to the fact that the modern concept of agricultural reform was very much widespread in those areas where the landowning elites were more involved in the world market, similarly to the slaveholding elites of the Americas, through their production of particular crops. At the same time, though, the European elites were also similarly fiercely committed, despite their paternalistic ideas, to keeping the status quo in their relations with the nominally free peasants.

As Marta Petrusewicz has shown, in nineteenth-century agrarian Europe, thanks to a very active public opinion, and to outstanding figures of agricultural reformers and political economists, the modern science of improving agricultural performance through agronomy was extremely popular among more economically than socially progressive landowners of several regions. This was especially the case of those landowners who lived in the Mediterranean commodity frontiers located in the European peripheries of southern Italy and Spain.[43]

In the nineteenth-century Mediterranean regions, agricultural modernization and agronomy were typically associated with horticulture, and therefore with the production of commercial agricultural items, such as citrus, olive oil, and wine. In southern Italy and Spain, not only scientific methods of farming, but also, to a lesser extent, technological innovation and debates tended to be particularly commonplace among those entrepreneurial landowners who engaged in horticultural cash-crop production. In nineteenth-century southern Italy, entrepreneurial landowners usually engaged in either horticulture or arboriculture (tree cultivation) on the coastal areas stretching from Campania to Sicily. They often kept citrus groves and vineyards, which embodied the very idea of agricultural progress, because of the high value of their products on the market. Progressive landowners participated in lively debates on scientific methods of farming in government-sponsored institutions called Economic Societies of the Provinces, or in private institutes—such as the *Istituto agrario* (Agrarian Institute) in Sicily—which gathered the most advanced agricultural reformers in different regions and published agricultural proceedings. An example of these proceedings from the 1850s is the *Annali di agricoltura siciliana* (Annals of Sicilian Agriculture), where examples of model landowners and model farms utilizing scientific techniques of cultivation abounded.[44]

Similarly, in nineteenth-century Spain, the areas of commercial agriculture on the coast stretching from Catalonia to Cadiz were the realm of horticulture and were grown with cash crops, among which particularly important were Valencian oranges and Catalonian wine grapes. On one hand, Spanish entrepreneurial attitudes tended to be associated with this type of commercial agriculture. On the other hand, agrarian reformers attempted to prompt the older landowning class, which lived off the rent of the grain-growing estates, to engage in agricultural modernization through the conversion of some of the land to horticulture. This feature shows particularly well in studies of the cultural and scientific press of the southern Spanish region of Murcia, which was *latifundia*-based. This was a region in which nineteenth-century agrarian reformers supported the necessity of modernization specifically through the implementation of chemical fertilizers and the renovation of traditional agricultural industries, such as silk. At the same time, progressive Spanish landowners gathered in regional institutions, such as the Cantabrian Economic Society, and also debated on prestigious economic journals, such as, from the 1830s, the Catalan-based *Revista de agricoltura pratica* (Review of Practical Agriculture), on which they engaged in discussions over what constituted modern agriculture.[45]

Though keeping in mind the fundamental differences between slave and free societies and forms of labor, we cannot help but notice the remarkable similarity between issues related to agricultural modernization among the slaveholding elites of the antebellum U.S. South and of the other regions interested by the phenomenon of the second slavery in the Americas and issues discussed by the most progressive members of the landowning elites of Italy's *Mezzogiorno* and Spain. If we also take into account the progressive landowning elites of Russia and Prussia, we can see that the movement toward agricultural modernization was a phenomenon spread among reforming slaveholders and landowners throughout the commodity frontiers located in the peripheries of the Euro-American world. Without going as far as arguing that all these Euro-American reforming elites read, or had heard of, the same agronomic studies by well-known German, English, and French luminaries (since this would need to be properly demonstrated), there is enough here to make us wonder what insights a full-fledged comparative study focusing on the nineteenth-century culture of Euro-American agrarian modernization would really yield.

American Second Slavery and European Second Serfdom and Sharecropping as Labor Systems

In comparison with the Atlantic plantation zone in the Americas, the large landed estate areas in eastern and southern Europe had several features that conditioned the structural organization and hierarchy of management of large landed properties in different ways from those of the New World plantations under the second slavery. In both cases, though, the influence of the restructuring of the nineteenth-century world market was far-reaching. In the case of plantations in the Americas under the second slavery, the very existence of highly organized and hierarchically managed agricultural units with slave labor in the cotton South, in sugar-producing Louisiana and Cuba, and in coffee-producing Brazil was due to the market demand for certain crops. In the case of landed estates in Russia, southern Italy, and Spain, the influence of market demand on organization and hierarchy of management was less obvious, simply because it affected mostly older systems of landed property worked through either free or unfree labor, forcing them to adjust. More important, the market's influence affected especially the areas of commercial agriculture—which produced grain, citrus, or wine—where direct management by progressive landowners, either from older or newer agrarian classes, was commonplace.

Looking back at the effects that the new nineteenth-century world market demand had on the different systems of labor used in the Atlantic plantation zone and large landed estate areas of eastern and southern Europe, it might be useful to not think of the different modes of agricultural work as dependent on absolute categories such as freedom and unfreedom. Rather, we might think of them as different labor systems on a continuum scale. At one end of this scale, we might place the types of gang systems and the mixed gang/ task system found under the second slavery in the U.S. South, Brazil, and Cuba as the most extreme forms of exploitation of both persons and labor. At the other end of the scale, we could put sharecropping as the least extreme form of exploitation, since it affected only contractual relations. In the middle of this continuum scale, we would place intermediate forms of exploitation of agrarian workers, such as the less exploitative *obrok* and the more exploitative *barshchina* types of *corvèe* present in Russian serfdom, the type of temporary indentured servitude suffered by the Chinese *coolie* labor employed in Cuba, and the more exploitative types of tenant and sharecropper contracts implemented in southern Italy and Spain.

Arguably, this would be a way to place both American slavery and slavery in the Americas in a much wider agrarian context, to which they both belonged as preeminently systems of agricultural labor related to the restructuring of the nineteenth-century world economy. At the same time, this diagram would also clearly show that, depending on the labor system, exploitation of nineteenth-century agrarian laborers occurred according to different features.

In systems of slave agrarian labor—as in the United States, Brazil, and Cuba—slaveholders responded to the market demand by applying additional pressure on regimentation and pace of labor. They changed the shape of labor management, taking the maximum advantage of the unfree condition of their workforce to obtain maximum profit from their agricultural business.

In milder systems of unfree agrarian labor, such as Russian serfdom, landowners could choose to extract the maximum amount of labor from the workers, with the corollary of strict management and supervision, at times of need, in order to increase production output in response to the world-market demand, as in the case of the Russian *barshchina* system and of the Chinese *coolie* labor in Cuba. Alternatively, landowners could extract simply monetary compensations, as in the case of the Russian *obrok* system, if they had little interest in engagement with the world market.

Finally, in systems of free agrarian labor, as in southern Italy and Spain, landowners could not force peasants either in relation to labor management or monetary compensation. However, they could still exploit them with unequal sharecropping contracts through which they might extract a large amount of labor over the course of several years, without allowing them to improve their condition of poverty. Exceptionally, though, sharecropping contracts could lead to equitable solutions in agrarian relations, as in the case of the Catalonian *rabassa morta.*

Arguably, among the different areas of possible comparison in relation to the changes that occurred on the plantations of the Atlantic world and on the landed estates of the eastern and southern European peripheries, agricultural modernization is the one that offers more possibilities in terms of exploring both direct and indirect links between the two regions. This is because the drive toward scientific agriculture and the increasing importance of the doctrine and practice of agronomy are themes that we can easily find among all the landowning classes that lived in the peripheral regions of the nineteenth-century world economy.

Implementing modernization—through the employment of scientific techniques of cultivation, through the stepping up of cash-crop production, through technological innovation, and through improvement of the transportation systems, and therefore through better market integration—was effectively the common way in which peripheral agrarian elites, on a global scale, chose to respond to the nineteenth-century world-market demand for ever-increasing and efficient production of old or new agricultural items. Since, despite the difference in the degree and nature of engagement in the world market, these themes are similar and comparable among agrarian elites throughout the peripheries of the Euro-American world—whether these elites were slaveholders, serfowners, or landed proprietors—we must conclude that this common entrepreneurial culture originated from a series of global transfers of ideas of modernization, not just within the Americas and Europe, but also between the two continents. However, only a detailed study of both the modes and the timing of the occurrence of ideas related to these global transfers will help us understand exactly when and how the latter actually took place in the course of the nineteenth century.

Chapter Five

ABOLITIONISM AND NATIONALISM ON THE TWO SIDES OF THE ATLANTIC

Within the contours of the now-established discipline of Atlantic history, scholars have treated the movements for the abolition of slavery that arose in different countries at different times between the 1820s and the 1840s as mostly discrete entities, with little relation to one another, unless they were part of the great Anglo-American cultural sphere.

In a recent synthesis in transnational perspective, Daniel Rodgers has claimed that all the great nineteenth-century reform movements in the United States were related to equally, if not more, important movements happening elsewhere. However, despite these very promising premises, what Rodgers really meant was that "[U.S.] Northern antislavery activists turned to England in search of connections, prestige, funds, and arguments." In other words, as a number of studies has shown, the true source of inspiration for American abolitionists such as William Lloyd Garrison was "William Wilberforce's and Joseph Sturge's England." In fact, as Robin Blackburn has remarked in relation to the connections and similarities between abolitionism in Britain and America, "While there were important differences between British and U.S. abolitionists, both of them operated in societies experiencing the disruption and class struggles of early industrialization."[1]

There are still very few studies, though, that put in relation British and American abolitionism with equally radical activities opposing slavery that arose in Ireland or France, for example. And yet, abolitionism was genuinely both transatlantic and intercontinental, and the frequent contacts between all the major figures of abolitionism, who mostly participated in the 1840 World's Anti-Slavery Convention in London, testify to that.[2]

Even less known and studied is the fact that the international dimension of those who opposed slavery did not confine itself within the abolitionist circles. Rather, it often embraced other progressive causes, including women's rights as well as nationalism intended as a self-justified struggle against national oppression—on both sides of the Atlantic.[3]

For true nineteenth-century progressives—individuals such as American William Lloyd Garrison, English William Henry Ashurst, Irish Daniel O'Connell, and Italian Giuseppe Mazzini—the fight against slavery was part of a much larger struggle that humankind had engaged in on its path to progress. This struggle was to lead to the eventual victory over all retrogressive institutions throughout the Atlantic world.

In the nineteenth century, the ideologies at the heart of antislavery, particularly its radical variant, abolitionism, and nationalism came to be linked by a strong belief in progress. As a number of important studies—first and foremost those by David Brion Davis—have shown, since the second half of the eighteenth century, slavery had become, in the minds of many intellectuals and in the public opinion, a backward, barbaric institution. As a consequence, emancipation and abolition stood as the very symbols of progress. Significantly, around the same time, nationalism also had started to be linked to the idea of progress, when the idea of the modern nation—and with it, the modern notion of citizenship rights derived from the Enlightenment—had spread throughout Europe as a consequence of the French Revolution.[4]

After hailing the birth of Euro-American liberal nationalism during the revolutionary period, and then the birth of independent nations in the area of the former Spanish-American Empire, in the first half of the nineteenth century, Atlantic progressives stood at the forefront of the second abolitionist wave, or the *new abolitionism.* This was a radical movement that, from the late 1820s onward, demanded immediate slave emancipation. Important victories in this sense were being obtained with the abolition of slavery in several of the new Latin American countries—from Chile (1823) to the Central American federation (1824) to Mexico (1829)—and with the abolition of Brazil's internal slave trade in 1831.[5]

In Latin America, the movement for the abolition of slavery had received an initial boost when the "liberators" Simon Bolìvar and José de San Martìn promised freedom to all the slaves who joined them in the struggle against the Spanish Empire. After national independence had been achieved, especially the states with smaller slave populations proceeded to abolish slavery altogether, whether with or without compensation for slaveholders. As a result, by then, a link between nationalism and abolitionism was forged in the minds of many people. This link would be constantly present, consciously or subconsciously, in the ideas that characterized the different Atlantic progressives.

While aware of these crucial developments, in the 1820s and 1830s, Atlantic progressives focused specifically on the struggles for the immediate abolition of slavery in the British Empire and the United States, all the while supporting in "thought and action" (Mazzini's famous slogan) the efforts at building liberal nations that were taking place in different parts of Europe. Thinkers and activists such as William Lloyd Garrison, William Henry Ashurst, Daniel O'Connell, and Giuseppe Mazzini were Atlantic progressives whom we may specifically qualify as nineteenth-century abolitionists, nineteenth-century nationalists, or both. They believed that the twin struggles against slavery and against national oppression led humankind along an increasingly progressive path. National independence, political self-determination, and freedom from oppression were all linked together in a great struggle whose aim was the progression of humankind toward a new, improved era based on the principles of liberty and justice.[6]

Put simply, for Atlantic progressives, there could be no real freedom—whether in the form of the abolition of slavery or in the form of national liberation—without equality. Thus, until that point of comprehensive freedom, humankind could not be said to be walking on its progressive path.

ATLANTIC ABOLITIONISM

As Carl Guarneri has noted, "antislavery agitators, like revivalist preachers before them, followed a northern transatlantic circuit, shuttling frequently between Britain and the United States to whip up popular support." In other words, nineteenth-century American and British radical antislavery advocates, from William Lloyd Garrison to John Sturge, as supporters of the new abolitionism, were ideal members of a sort of "abolitionist international." Both Irish and French antislavery activists belonged to this group as well.[7]

This movement's roots lay deep into the Age of Revolutions. Specifically, there was a strong link with the transformations brought to the Atlantic world by the spread of the notions of human rights and democratic equality from the United States and France, passing through Haiti. The momentum had waned after reaction had followed revolutionary change on both sides of the ocean, and after the Atlantic economy had instilled new life into the second slavery after 1800 (as discussed in previous chapters).

On one hand, though, as Christopher Schmidt-Nowara has noted, "the specter of the Haitian revolution haunted plantation belts" of the Americas with the nightmare scenario of a bloody end to slavery. On the other hand, abolitionists were also quick in regrouping and increasing their pressure over the issue of slave emancipation on various national governments. Thus, we should see the succession of provisions for the abolition of slavery in the first half of the nineteenth century—first in Britain, between 1833 and 1838, and then in France, in 1848—partly as the crowning achievement of a single, Atlantic, radical abolitionist movement, with different national strands or varieties.[8]

American Abolitionism until 1840

In the 1820s, in the United States, most white antislavery advocates gathered in the American Colonization Society (founded in 1816) and supported plans for returning slaves to Africa after gradual emancipation had been achieved. However, African American activists pursued their own radical new abolitionist agenda, which had a distinctive Atlantic dimension. Thus, when in 1829, David Walker launched his famous *Appeal to the Colored Citizens of the World,* in which he denounced colonization and incited the slaves to take arms and rebel against their masters, he imagined himself to be speaking to all slaves of African descent in the Americas, and beyond.[9]

Walker's crucial influence on William Lloyd Garrison's enunciation of "immediate, unconditional, uncompensated emancipation," from the very first issue of his radical antislavery journal *The Liberator* in January 1831—the conventional date of the beginning of the new abolitionism in the United States—is now well established in the historiography, and maybe Walker also inspired Nat Turner and his revolt in Virginia in 1831. Equally crucial is Walker's influence on Garrison's growing understanding of the Atlantic dimensions of the abolitionist struggle. Beginning with the 1832 foundation of the New England Anti-Slavery Society (NEASS), in the decade of the 1830s, under Garrison's guidance, abolitionism in the United States grew to become a national movement, building at the same time close transatlantic links with supporters from other countries.[10]

In 1833, 62 representatives of the three main abolitionist groups based in the United States met in Philadelphia and founded the American Anti-Slavery Society (AASS), the first national organization whose aim was the immediate emancipation of slaves in the U.S. South. Its "Declaration of Sentiments," written by Garrison, endorsed the Christian attitude of nonresistance and condemned slavery as a violation of the principles stated in the Declaration of Independence and as a sin in the eyes of God. At the same time, the "Declaration of Sentiments" urged abolitionists to persuade citizens of the northern states to endorse the abolition of slavery through the tactic of "moral suasion" and the spread of abolitionist literature.[11]

A year after its foundation, the AASS began publication of its own periodical, *The Emancipator,* which was circulated nationally. Several other periodicals, such as *the Anti-Slavery Reporter* and *The Anti-Slavery Standard,* joined *The Emancipator* in an impressive flow of publications. These also included books and thousands of pamphlets aimed at showing Americans, especially northerners, the evils of southern slavery. As Thomas Bender has recently argued, "for the South, improved and cheaper mails brought closer the threat of northern ideas, particularly those of the Abolitionists," while at the same time creating a "heightened sense of territoriality, of unified and uniformed national territories … evident throughout the Atlantic World."[12]

And within this Atlantic world, characterized by an increasing importance of both abolitionism and nationalism, Garrison and the Garrisonians (his supporters in the AASS, mostly from the original New England group) moved freely back and forth, as Frederick Douglass's trip to Britain and Ireland in 1845 through 1847 testifies. In fact, as Ian Tyrrell reminds us, "American Abolitionists frequently toured Europe to draw moral and financial support." Garrison himself went to Britain and Ireland three times in the 1830s and 1840s—in 1833, 1840, and 1846—"each time stirring controversies."[13]

Garrisonians thought that reform was needed in several spheres to bring about a new society, first and foremost in the United States, but also in the world at large. Hence, the sentence on top of every issue of *The Liberator* was "Our country is the world, our countrymen are mankind." Consequently, Garrisonians did not restrict themselves to fighting against slavery, but became involved in other movements for social reform in America. They also actively supported progressive national movements in Europe, and they maintained a general attitude of questioning accepted authority.[14]

Unsurprisingly, Garrisonians also welcomed women in the abolitionist movement and encouraged them to join the AASS. Many of them became actively involved in the society's activities.

By the later 1830s, women had started merging the fight against slavery with the struggle for female rights and equality. Garrisonians, with their broader Atlantic view, supported women in a struggle that was fast becoming transnational in its modes and links, as part of the necessary moral revolution of society. However, anti-Garrisonians, who were mainly abolitionists based in New York with a much narrower view, considered the movement for women's rights extraneous to abolitionism and unnecessary. Finally, at the 1840 meeting of the AASS—the same year as the first World's Anti-Slavery Convention—anti-Garrisonians were defeated in the elections of the executive committee by the Garrisonian faction, which succeeded in having a woman elected for the first time.[15]

Most of the anti-Garrisonians left and went to fill the ranks of the newly founded American and Foreign Anti-Slavery Society (AFASS), while the Garrisonians remained in control of the original AASS. Even though the AASS diminished in numbers, for the next 25 years, it remained a far more radical and transnational organization, both in its principles and in its links with various Atlantic progressives.[16]

British Abolitionism and the 1833 Emancipation Act

Within the contours of Atlantic abolitionism, there were close links between Britain and the United States in the first half of the nineteenth century. For this reason, it is possible to see clear parallels in the development of antislavery sentiment and achievements in the two countries. This is apparent from the abolition of the Atlantic slave trade (in 1807 in Britain, and in 1808 in the United States) to the making of societies invoking colonization, or the return of blacks to Africa, and gradual abolition in the 1810s and 1820s, to the rise of much more radical institutions demanding the immediate abolition of slavery.[17]

After winning a great victory with the 1807 abolition of the Atlantic slave trade, most antislavery advocates in Britain were convinced that its effects would lead to a general amelioration of the life of the slaves in the British colonies in the West Indies. However, this proved not to be the case. Thus, several of them gathered in 1823 to form a Society for the Mitigation and Gradual Abolition of Slavery throughout the British Dominions, or the Anti-Slavery Society (ASS). Among the members were veteran antislavery activists Thomas Clarkson and William Wilberforce, and the latter's acknowledged successor Thomas Fowell Buxton. The ASS's objectives were primarily the protection of slaves from mistreatment, and, similarly to the American Colonization Society on the other side of the Atlantic, the achievement of gradual emancipation.[18]

Though the ASS was instrumental in forcing the British Parliament to better the conditions of slaves through measures that promoted religious instruction, removed obstacles to manumission, and, above all, succeeded in banning the flogging of women, its limited achievements, due mainly to the planters' resistance, could not prevent the periodical occurrence of slave rebellions. Two large revolts—the 1823 one in Demerara and the 1831 to 1832 Baptist War in Jamaica—framed the period of transition of the British antislavery movement from gradualism to immediatism. This was also following the publication of Elizabeth Heyrick's pamphlet *Immediate, not Gradual Abolition* in 1824.[19]

In 1830 to 1831, the London group of the ASS reorganized itself and broke away, embracing the new abolitionism, and establishing the radical Agency Committee. Similar to the AASS in the United States, the committee's goal was the dissemination of abolitionist literature and ideas, and the signing of petitions in order to persuade the people to endorse, in the words of Howard Temperley, the "immediate and unconditional emancipation of all the slaves in Britain's overseas possession." [20]

Partly as a result of this flurry of activity, partly fearing more massive slave insurrection, and partly influenced by the increasingly popular argument on the moral and economic superiority of free labor over slavery, in 1833, Parliament passed the Emancipation Act, which freed almost 800,000 slaves in all the British colonies by August 1, 1834. The Act compensated slaveholders for their loss, stating that ex-slaves had to undergo an unpaid "apprenticeship" period originally of 12, then 6, and finally 4 years, so that actual emancipation really occurred in 1838.[21]

According to David Brion Davis, "whatever compromises seemed necessary, Britons and many Americans hailed the Emancipation Act as one of the greatest humanitarian achievements in history." And, in truth, from an Atlantic abolitionist perspective, British slave emancipation—called "the mighty experiment"—had immense significance, especially considering, in the words of Deborah Bingham Van Broekhoven, that "the handful of American abolitionists ... praised Britain and its reformers, and for decades they continued to compare their own struggle with the swift end of slavery in the West Indies."[22]

Yet, even after the Emancipation Act was passed, the most radical abolitionist members of the Agency Committee regrouped in 1837 under the leadership of Joseph Sturge in the Central Negro Emancipation Committee. Their pressure on the British Parliament was crucial in reducing and then ending apprenticeship altogether.

In 1839, after winning the battle over apprenticeship, many individuals of the same group, still led by Sturge, founded the British and Foreign Anti-Slavery Society (BFASS), whose declared aim was to work for the eradication of slavery throughout the world. Interestingly, Sturge went on to "become a leader of the 'moral force' wing of Chartism in the 1840s," according to Robin Blackburn. At the same time, in the following years, the BFASS collaborated closely with antislavery activists in different countries, even though it had little success in fighting the Atlantic slave trade, especially in Brazil and Cuba, also as a result of the crucial role played by British companies and investments. Also, in truly cross-continental fashion, it organized the World's Anti-Slavery Conventions of 1840 and 1843, which were two major events in the story of Atlantic abolitionism and the transatlantic links between the United States and Europe.[23]

Irish Abolitionism until 1840

In the context of Atlantic abolitionism, Ireland might appear as somewhat peripheral, mainly because the fight against slavery did not affect an established system of economic interests to the same extent as in the United States, Britain, or France. Partly as a consequence of this, the antislavery movement in general was never as popular in Ireland as in Britain and the United States. As Maurice Bric has noted, the antislavery movement "was perceived [with some reason] by some Irish reformers as having not only a Protestant, but also a powerful anti-Catholic strain."[24]

Yet, Ireland has proven repeatedly, as a large recent historiography attests, an important case study in the story of transatlantic exchanges, given its position, culturally as much as geographically, in between the American and the European worlds. The story of Atlantic abolitionism can only confirm this view, since, also within this context, the Irish culture of reform of the first half of the nineteenth century—especially as expressed by the monumental historical figure of Daniel O'Connell—provides a crucial link between a well-established Anglo-American tradition of radical antislavery and an incipient tradition of European nationalism, or at the very least, of recognition of the rights of oppressed nationalities.[25]

In Ireland, the antislavery movement had been active since at least 1824, with the foundation of the Dublin Association for Endeavoring to Promote the Mitigation and Gradual Abolition of Slavery in the British Colonies. Five years later, Charles Orpen founded the most important Irish antislavery society, the Dublin-based Hibernian Negro's Friend Society (HNFS).[26]

The HNFS was sympathetic to Britain and dominated by Protestants. Even in 1830, Protestants and Quakers were a driving force behind Irish antislavery. This is especially apparent when we look at women's activism as expressed through the Hibernian Ladies Negro's Friend Society, founded in 1833, and the Dublin Ladies' Antislavery Society, founded in 1837. However, it is also true, as Nini Rodgers states, that "from 1801 to 1833, from the passing of the Irish Act of Union to the abolition of slavery throughout the British Empire, within the Westminster Parliament, the issues of Catholic Emancipation and African slavery were closely linked." Note that in 1823, the same year in which Thomas Fowell Buxton (an Englishman who had been educated at Trinity College Dublin) founded the Society for the Amelioration and Gradual Abolition of Slavery, O'Connell began the activities of the Catholic Association in Ireland.[27]

Largely as a result of O'Connell's efforts and his widespread popular support, in 1829, the British Parliament granted Catholic emancipation by passing an act that allowed Irish Catholics the right to be elected as Members of Parliament (MPs) without being forced to abjure their own confession. The importance of the act from the point of view of a study of Atlantic abolitionism is clear when we consider that, also as a consequence of the wave of reform that started with Catholic emancipation, the movement for the abolition of slavery in the British Empire reached its momentum in the early 1830s and led to the British Parliament's passing of the Emancipation Act of 1833.[28]

Throughout those four years—1829 through 1833—O'Connell had been at the very forefront of debates, both within and without Parliament, in which he supported slave emancipation wholeheartedly. After 1833, he was at the forefront of the struggle to end the apprenticeship system for ex-slaves. Thus, as a Catholic, O'Connell was instrumental in creating an alternative abolitionist tradition in Ireland. In this tradition, an embryonic form of Irish nationalism in its milder version of struggle for Catholic parliamentary representation was very much tied to the suppression of slavery as part of a general move toward progress through reform. To this end, not only did O'Connell place himself at the center of transatlantic abolitionist networks, becoming a close friend of William Lloyd Garrison, but he also overcame his own initial diffidence and allied with the Protestant and Quaker abolitionists of the HNFS, which in 1837 changed its name to Hibernian Anti-Slavery Society (HASS). By 1840, when the World's Anti-Slavery Convention took place in London, Irish abolitionists had come to recognize that they all were working toward the same goal.[29]

French Abolitionism to the 1848 Abolition of Slavery

From the perspective of Atlantic abolitionism, France provides an altogether different case, but nonetheless an important one. French abolitionism played a fundamental historical role, and one that provides crucial insights on its links within the Atlantic world.

The French abolitionist tradition, even more than the British one, stretched all the way back to the Age of Revolutions. Already in 1788, there existed a *Societé des Amis des Noirs* (Society of the Friends of the Blacks), which included prestigious members such as Mirabeau, Condorcet, and Lafayette.

The subsequent events in the French Revolution brought to power the Jacobins, who in 1794, decreed the emancipation of slaves in the French colonies, partly as a response to the Haitian Revolution, which led to the end of slavery in St. Domingue. The events also led to the foundation of the short-lived *Societé des Amis des Noirs e des Colonies* (Society of the Friends of the Blacks and of the Colonies, 1796–1799). But then, in 1802, Napoleon reinstated both the slave trade and slavery in the colonies.[30]

Thus, starting from France and expanding to the rest of the Atlantic world in the Age of Revolutions, a pattern developed in which the abolition of slavery became clearly linked to the construction of the nation, and specifically of a French democratic nation. Even though this link was forced underground during the Napoleonic and early Restoration period, it provided a lasting model for the national revolutions that led to the formation of abolitionist republics in Latin America in the 1810s and 1820s, and also for the national revolutions that would lead to democratic and antislavery governments in France and the rest of Europe in the 1830s and 1840s.[31]

Starting from the 1820s, in their opposition to slavery, both Protestant and Catholic French abolitionists received inspiration and support from Britain. In 1821 they founded the mild, liberal-oriented *Societé de la Morale Chretienne* (Society of Christian Morality), which counted among its members the future king, Louis Philippe. Besides opposing the reactionary French kings and governments in the name of the formation of a liberal nation, the society's activities relating to slavery were directed mainly to ameliorating the lives of the slaves and opposing the slave trade. However, its most radical members, especially Auguste de Stael-Holstein, also talked and wrote about the abolition of slavery.[32]

Eventually, in 1831, after the 1830 revolution and the coming to power of the liberal government headed by Louis Philippe, the slave trade was suppressed. In France, as in Britain, the abolitionist movement focused its efforts on the emancipation of all the slaves in the colonies. In this connection, it is fair to say that only through the perspective of Atlantic abolitionism is it possible to see that the effects of the British Emancipation Act of 1833 went well beyond the English-speaking world and provided a very large boost to antislavery activities in several other areas, including France.[33]

In 1834, shortly after the British Emancipation Act, encouraged by British antislavery activists, French abolitionists founded the *Societé Francaise pour l'Abolition de l'Esclavage* (French Society for the Abolition of Slavery), headed by Francois Isambert, Alexis de Tocqueville, and others. From the perspective of Atlantic abolitionism, the French Society was far more gradualist than others, especially its American counterpart the AASS, since the idea that slaves needed to be prepared for their freedom accompanied its request for emancipation. Still, it was thanks to the pressure brought by the French Society on Parliament that slave emancipation became, for the second time, a truly national issue.[34]

As in the United States, free black abolitionists, such as Cyrille Bissette, were far more radical and immediatist, as, indeed, were a minority of whites, such as Victor Schoelcher. At the same time, similarly to what happened in Britain, immediatist positions became increasingly popular in France, partly due to the fear produced by frequent slave revolts in the colonies. Delegates from the French Society also joined the international abolitionist movement participating at the World's Anti-Slavery Convention in London in 1840, where the predominant attitude was both abolitionist and immediatist. The French abolitionists in general began to embrace immediatism after 1845, and then more forcefully starting in 1847.[35]

Shortly afterward, in February 1848, a new revolution overthrew the July Monarchy of Louis Philippe and established a national French Republic. In that government sat immediatist Victor Schoelcher, who did not waste time and decreed emancipation almost immediately, freeing more than 235,000 slaves in the French colonies in the Caribbean.[36]

The abolition of slavery in the French colonies in 1848, in turn, led to similar measures taken by the Dutch and Danish governments for slave emancipation in their New World colonies. Thus, 1848 truly stands as a significant date at the conclusion of an antislavery momentum caused by the rise of new abolitionist movements in the second half of the 1820s.

In a relatively short time, the movements, located in the main countries that controlled the slavery business with hundreds of thousands of people in bondage, had succeeded in putting in practice agendas that were mostly immediatist, exercising pressures on their respective national governments for immediate emancipation. Moreover, the abolitionist movements had become part of a wider Atlantic abolitionism, whose distant origins were in the Age of Revolutions. Now, through transatlantic contacts and connections—which reached their peak at the time of the World's Anti-Slavery Convention of 1840—abolitionists were able to spread the new immediatist version of abolitionism from the English-speaking to the French-speaking areas of the Atlantic world and beyond.

As a result of both these transatlantic contacts and the actions of the slaves, in the words of Thomas Benjamin, "from the early 1830s to the late 1840s most of the slave systems of the Caribbean, the center of the New World plantation complex, were destroyed by European reformers and by the slaves themselves." Equally important was the fact that, in the process of the destruction of the slave systems, the national governments acted under the pressure of national antislavery organizations. These organizations, even though part of an Atlantic abolitionist movement, acted in the name of overall programs of national reforms, thus linking once more—as had already happened during the Age of Revolutions—the concepts embodied in the creation of democratic nation-states and in the abolition of slavery.[37]

ATLANTIC PROGRESSIVES AND THEIR CAUSES

Within the context of Atlantic abolitionism, we now situate the "thought and action" of the four Atlantic progressives referenced earlier in the chapter: first and foremost William Lloyd Garrison, and then William Henry Ashurst, Daniel O'Connell, and Giuseppe Mazzini. What linked their activities and led them, in three cases out of four, to become acquainted with one another was their shared belief in the centrality of the radical fight for the abolition of slavery as an indispensable condition for humankind's march toward progress. Thus, after emancipation had been achieved in the British Empire in 1834, they all participated in or expressed their support for the international efforts of the World's Anti-Slavery Conventions. These conventions were directed toward the abolition of slavery worldwide and especially in the United States, where William Lloyd Garrison headed an increasingly radical and transnational abolitionist movement.

At the same time, during the 1830s and 1840s, as national revolutions sprang up all over Europe, Atlantic progressives, already aware of the connection between radical antislavery and democratic nationalism, actively supported, in different ways and degrees, the various struggles for national self-determination. The most extreme example of this was represented by Giuseppe Mazzini's activities for the liberation of the Italian nation.

William Lloyd Garrison

Among Atlantic progressives, William Lloyd Garrison, born and raised in New England, was certainly the one who provided the most indispensable link between the different strands and varieties of Atlantic abolitionism and European democratic nationalism. In 1831, under the influence of David Walker and the black Atlantic abolitionists, Garrison began, with unequivocal words published in the first issue of *The Liberator* ("I shall strenuously contend for the immediate enfranchisement of our slave population"), the most influential immediatist movement in the United States.[38]

Garrison's main reason for demanding immediate emancipation—notably that slavery was a violation of the principles written in the Declaration of Independence—might have seemed narrow and circumscribed to the American case. However, in beginning an immediatist tradition in the United States, Garrison tightened the connections between American and British antislavery, creating the preconditions for an essential link between radical abolitionists, most of whom were immediatist, in both countries, as well as in Ireland and later in France. In this sense, Garrison's contribution to Atlantic abolitionism is incalculable.[39]

On the other hand, it was Garrison's own increasingly universal vision of the problem of slavery that led him to broaden his view of immediate emancipation and consider it part of a general reform. This reform was first of American society, and then of the world at large, and it was directed toward the elimination of all the obstacles to humankind's march toward progress. These were obstacles that Garrison, being highly religious, saw first and foremost as sins in the eyes of God.[40]

We can see the seeds of this attitude already in the "Declaration of Sentiments," the programmatic document with which Garrison stated the aims of the AASS in 1833. In this document, the fight for the abolition of slavery in the United States also assumed a universal value, since its ultimate success would have been a "triumph of JUSTICE, LIBERTY, and HUMANITY." It was precisely this universal attitude that led Garrison and the Garrisonians to embrace an increasingly broader and more radical view of immediatist abolitionism. This was a view that eventually caused the 1840 separation, leaving Garrison in control of the AASS.[41]

After the 1840 separation, Garrison tightened his links with Atlantic abolitionism, becoming a central figure in it, especially by participating at the World's Anti-Slavery Convention in London in 1840. On the other hand, Garrison became an increasingly vocal supporter for other, equally radical reform movements, which he, as a now mature Atlantic progressive, rightly saw as tightly linked to the cause of Atlantic abolitionism: the women's rights movement and democratic nationalism.[42]

Thus, Garrison became involved with the activities of the most important women's rights activists of his generation, several of whom he had welcomed in the AASS. More important, he became acquainted with some of the most famous nationalist politicians and activists of the time, notably Daniel O'Connell and Giuseppe Mazzini.[43]

Significantly, Garrison first became acquainted with O'Connell and Mazzini as a result of their support for an absolute uncompromising position, close to Garrison's own, on the immediate emancipation of the slave population in the United States. Their involvement with each other is an important testimony to how the network of Atlantic abolitionism gathered activists from different countries and backgrounds fighting for similar ideals. In the course of the 1840s, Garrison supported O'Connell's positions on the Irish question almost until the latter's death in 1847. He also strongly supported Mazzini's democratic nationalist activities in Italy, especially at the time of the 1849 Roman Republic, when Garrison went as far as publishing in *The*

Liberator periodic reports by William Henry Ashurst on Mazzini's democratic achievements in Rome.[44]

William Henry Ashurst

It was hardly a coincidence that Garrison's correspondent from Mazzini's Roman Republic for *The Liberator* was William Henry Ashurst. Born in London and trained as a lawyer, during the 1830s Ashurst had become a convinced follower of the utopian socialist ideas of Robert Owen. Through the publication of different journals, notably *The Reasoner* and *The Spirit of the Age,* Ashurst had become known to British society as a political activist and a radical. In the articles published in the pages of *The Reasoner,* he first used his pseudonym Edward Search, and he continued to use it throughout his life while following and reporting on different radical activities.[45]

True to his statement, as remembered by his daughter Emily Venturi (author of a brief biography of her father), that "agitation is for the people that troubling of the water which produces health and vigor," Ashurst became involved in a variety of radical causes, which made him the true quintessential Atlantic progressive. As a result, we can hardly be surprised that, in the coverage of the time, according to Matthew Lee, Ashurst's involvement in "feminism, anti-slavery, European (and especially Italian) nationalism and republicanism, secularism, and administrative reform paint the picture of a uniquely energetic and vigorous man."[46]

For Ashurst, antislavery was one of many causes that he saw connected in a broad crusade for progress and against tyranny and oppression. And yet, as in every other cause he embraced, he was radical also in supporting antislavery. Thus, it is no wonder that, when Ashurst met William Lloyd Garrison at the 1840 World's Anti-Slavery Convention, they immediately became friends. Ashurst and Garrison began a long collaboration, expressed mainly through regular contributions to *The Liberator,* as much on radical abolitionism as on women's rights, both subjects for which Ashurst was constantly campaigning.

Thus, through his own particular interests, Ashurst became an indispensable link between Atlantic abolitionism and other forms of radical activity, as well as between Atlantic progressives living on the two sides of the ocean and involved in parallel and connected causes, from American abolitionism to women's rights, to European democratic nationalism. For the best part of 20 years, Ashurst's house in Muswell Hill in London became a meeting point for an ever-larger group of radical activists, nicknamed the "Muswell Hill Brigade," who gathered there on Sunday afternoons. One of those activists was Garrison whenever he was in London. After Ashurst's meeting with Giuseppe Mazzini in 1844, the Italian patriot also became a regular visitor.[47]

Eventually, in 1846, at Ashurst's house, Garrison and Mazzini first became acquainted. In that same year, Ashurst collaborated with Mazzini, becoming a founding member of the People's International League, an organization that

Mazzini created to carry on the program of his former Young Europe. Later on, Ashurst followed the Italian nationalist in the adventure of the rise and fall of the 1849 Roman Republic.[48]

During the days of the Roman Republic, Ashurst not only went along with Mazzini, but he also documented the events and dutifully reported back to Garrison, who published his regular articles (signed as Edward Search) on Italy's situation in *The Liberator.* Significantly, Ashurst died in 1853, while visiting Garrison and his fellow abolitionists in the United States. So, he functioned as a link between Atlantic abolitionism and European radicalism and democratic nationalism up until the end.[49]

Daniel O'Connell

Similarly to William Henry Ashurst, Daniel O'Connell was trained as a lawyer, but he was also Irish and a Catholic landowner. Both O'Connell's privileged background and his horror at the Jacobin excesses of the French Revolution ensured that he would maintain a clear hostility toward radical political action. At the same time, his readings in the formative years—which included Voltaire, Rousseau, and Tom Paine—led to his sincere and enduring espousal of liberal and progressive principles. Thus, according to Gearoid O'Tuathaigh, "civil and religious equality, freedom of conscience, and the extension of individual liberty became the basis of his [O'Connell's] philosophical and political position."[50]

After the 1801 Act of Union of Britain and Ireland failed to abolish the legal discrimination of the Irish Catholics—barred both from Parliament and from the State's high offices—O'Connell threw himself body and soul into the fight for Catholic emancipation, eventually founding the Catholic Association in 1823. O'Connell's nature as an Atlantic progressive showed in his political innovations, which became a model to follow for subsequent Atlantic reform movements, particularly his campaign initiatives to mobilize the Irish people on an unprecedented mass scale.[51]

Between 1824 and 1828, through these mass campaigns, O'Connell pressurized the public opinion, eventually gaining a parliamentary appointment, and achieving a status of European reformer and advocate of national rights. O'Connell's status achieved Atlantic resonance with the British Parliament's 1829 passing of the Catholic Relief Emancipation Bill.[52]

Once in Parliament, throughout the 1830s and 1840s, O'Connell continued to fight for the improvement of conditions of the Irish Catholic population, notably through his long-term campaign for the repeal of the 1801 Act of Union. He proved until the end to be a progressive reformer with the interests of the Irish nation at heart. He was not a revolutionary nationalist narrowly focused on the objective of national self-determination, though, as were the members of the Young Ireland movement (1842–1848).[53]

What made O'Connell an Atlantic progressive was a combination of factors that showed how he, even more than William Henry Ashurst, proved to be an indispensable link between different causes that Atlantic progressives were pursuing on both sides of the ocean. Aside from the fact that his campaign for Catholic emancipation provided a model for various reform movements, and especially for Atlantic abolitionism, O'Connell himself proved for Atlantic progressives a model to imitate. He was nicknamed "The Liberator," like Bolivar and also the name that Garrison gave his abolitionist paper. O'Connell's role model essentially stemmed from the resonance and success of his own involvement in progressive causes directed toward the amelioration of the conditions of humankind, whether the latter was represented by discriminated Irish Catholics or by exploited American slaves.[54]

Throughout the 1830s and 1840s, until his death in 1847, O'Connell became personally involved in the parliamentary debates such as the ones that first brought about British emancipation, and then the end of the apprenticeship system. As a consequence, he became a major figure in Atlantic abolitionism.[55]

O'Connell represented Ireland and the HASS at the World's Anti-Slavery Convention in London in 1840, where he became acquainted with William Lloyd Garrison. Significantly, it did not take long for O'Connell to realize how much the two of them had in common. Not only were they both Atlantic progressives, but they also were both committed to achieve radical objectives through peaceful, democratic means. Ultimately O'Connell provided Atlantic progressives such as Garrison and Ashurst, whom he also knew, with an indispensable link not only between two different types of progressive causes in the Atlantic world—European (specifically Irish) nationalism and Anglo-American abolitionism—but also between radical reform and the democratic means to achieve it.[56]

Giuseppe Mazzini

Giuseppe Mazzini had many features in common with Garrison, Ashurst, and O'Connell, although he maintained a long-standing friendship only with the former two. And, even though the nature of his Atlantic progressivism was demonstrated through his involvement in a variety of different causes, like the other three individuals mentioned, Mazzini came to be identified very soon with the cause of Italian and European democratic nationalism.

The same age as Garrison, Mazzini lived his early years under the repressive regime of the Kingdom of Sardinia, in northwest Italy. He had seen the Italian patriots' various attempts at overthrowing despotic regimes and at forming an Italian nation free from foreign rule crushed one after the other. After an initial spell with the *Carbonari* secret society and the failure of his own revolutionary attempt, Mazzini went into exile in Marseille. From there, in 1831, he founded Young Italy, and the following year, he began publishing his paper *La Giovine Italia.* Much as Garrison's *The Liberator* did with the new abolitionist movement in those same years, Mazzini's *La Giovine Italia* spelled out unequivocally the program of a new type of commitment to the national cause. This was a commitment to the creation of an Italian "republic, one, indivisible" and to the "abolition of every aristocracy and privilege." Mazzini, thus, created a new tradition of democratic nationalism—one that, as in the case of O'Connell's nationalist movement for parliamentary reform, was going to provide an enormously successful model in Europe and beyond.[57]

From the beginning, Mazzini's activities had a symbolic and lasting meaning that went well beyond the narrow confines of Italian nationalism. Then, in 1834, he founded Young Europe, and attempted to extend—not just symbolically, but also practically—his Italian model to the struggles for the recognition of oppressed nationalities that were taking place in different parts of Europe, from Germany to Poland. From then on, Mazzini stopped being simply an Italian nationalist and became the most prominent European democratic nationalist. He was an intellectual and agitator whose grandiose vision for the future was the coordination of all democratic movements in Europe to create a continent of nations free from despotic regimes. Despite the repeated failures of all his programs and his willingness to sacrifice young lives through revolutionary violence, in his years of exile—which in the 1830s brought him to his long-term destination in London—Mazzini found himself surrounded by the admiration of European and American activists for the greatness of his vision and for his absolute ethical commitment.[58]

Even though Mazzini only supported, and did not participate in, the 1840 World's Anti-Slavery Convention, it was still in London that he met Ashurst in 1844 and Garrison in 1846. In both cases, the meeting signaled the beginning of a lifetime friendship.

By the 1840s, Mazzini was well acquainted with the world of British and American reform, and he had become a champion of radical causes. He was particularly well known among the main proponents of the Chartist Movement, and their working-class radicalism influenced him deeply.

Particularly through the British radical circles, Mazzini came in contact with Atlantic abolitionism. Through his friendship with Garrison, he embraced it wholeheartedly, becoming as prominent a European abolitionist and true Atlantic progressive as O'Connell had been. Thus, while Mazzini continued to operate for the liberation of Italy and other oppressed nationalities in Europe, he did not refrain from speaking out against slavery in radical abolitionist tones, making comparisons between the "black slaves" in the United States and the "white slaves" in Europe. Even though it is true that, in the case of Ireland, he did not support the repeal of the 1801 Act of Union, there is no doubt that Mazzini effectively provided, even more than O'Connell, an effective link between Atlantic abolitionism and European nationalism.[59]

Ultimately, both American abolitionists and British reformers—first and foremost Garrison and Ashurst—showed that they understood the significance of that connection when they demonstrated their unwavering support for Mazzini's failed attempt at establishing the embryo of an Italian democratic republic in Rome in 1849.

Each in his own way, the four characters of William Lloyd Garrison, William Henry Ashurst, Daniel O'Connell, and Giuseppe Mazzini demonstrate the features of mid-nineteenth-century Atlantic progressives. The basic feature they all had in common was their commitment, in varying degrees, to a number of different causes that were directed toward the amelioration of large sections, or even the whole, of humankind. All four men were involved in movements that addressed in the most radical ways the pressing problems related to inequality of race, gender, class, and national oppression: antislavery, women's rights, abolition of privileges, and democratic nationalism.

However, we can also say that the involvement of these four Atlantic progressives in these different movements functioned at two different levels. On one hand, it was related to the particular problems that the country, or "imagined" country, that they belonged to faced, and thus it focused on the one movement that addressed the most pressing national problem: abolitionism in the United States, abolitionism and reform in Britain, Catholic emancipation in Ireland, and nationalism in Italy. On the other hand, the Atlantic progressives' international dimension, as a result of the existence of established networks within Atlantic abolitionism and European nationalism, allowed them to see the specific problems of their own nations as parts of a broader set of problems that the entirety of humankind faced. Ultimately, they collaborated in progressive causes because they all saw those causes as connected together in a worldwide effort to uplift humankind and put it on its rightful progressive path.

ATLANTIC ABOLITIONISM AND EUROPEAN NATIONALISM

The strength of the connections between the different abolitionist movements into one large Atlantic abolitionism had its test at the World's Anti-Slavery Convention held in London in 1840. There, for the first time, delegates and representatives of the abolitionist movements of Britain, the United States, and France gathered together to discuss an agenda whose first objective was the abolition of slavery in the entire world. However, and despite different problems, the World's Anti-Slavery Convention ended up being much more than simply a celebration of Atlantic abolitionism.

On one hand, the issue of women's rights had an important part in the convention. On the other hand, the presence of reformers involved in many different progressive causes other than abolitionism broadened its scope and resonance. Thus, the fact that three out of the four Atlantic progressives we have analyzed—all world-renowned public figures—participated in the convention, while the fourth supported it, shows the strict link between Atlantic abolitionism and other reform movements, particularly European nationalism.

If 1840 was the peak of transatlantic activity and ardor for Atlantic abolitionism, 1848 was doubtless the peak of the continental embrace of European nationalism. In 1848, nationalist movements sprang up almost everywhere in Europe. Most of them were democratic and aimed at the establishment of republican nation-states.

The 1840 Anti-Slavery Convention and the 1848 nationalist movements, though occurring in different years, should be seen as parts of a common story, whose links are represented by Atlantic progressives. The Atlantic progressives' activities not only partly caused the events to happen, but also influenced them in different ways and degrees.

The 1840 World's Anti-Slavery Convention

Realizing the importance of showing the world the strength of connections of a united front of Atlantic abolitionism, in 1839, Joseph Sturge, head of the BFASS, issued a general call for delegates for a World's Anti-Slavery Convention to be held in London the following year. In the words of Douglas Maynard, "the calling of a 'General Anti-Slavery Convention' to meet in London in June 1840 represented an attempt to strengthen and enlarge the crusade against slavery by drawing together into one combined effort the abolitionist forces of the mid-nineteenth century."[60]

Although the proceedings list more than 500 delegates, the actual attendance numbered 409. Understandably, the majority of the delegates were British abolitionists. The number of delegates from Scotland, Ireland, and the British Caribbean was 50; 40 delegates were from the United States; 4 delegates were from France; and 1 delegate each came from Haiti, Spain, and Switzerland. The disproportionate number of British participants, even compared to the American delegation, has led some scholars to question the world significance and resonance of the convention. But the truth is that, from the perspective of Atlantic abolitionism, regardless of the actual numbers, the leading activists were all present, and they discussed and deliberated together for the first time, making decisions that would affect the course of the fight against slavery worldwide.[61]

The World's Anti-Slavery Convention lasted 13 days, from June 12 to June 25, 1840. In general, participants saw reports about slavery in different areas of the world. The convention also led to discussion of the best strategies to fight slavery. They decided these strategies should focus on an increased use of the press and the publication of pamphlets, together with the presentation of antislavery resolutions to the governments of those countries in Europe and the Americas that supported or allowed slavery.[62]

From the first day, the issue of the participation of women catalyzed attention, especially because of the number of female delegates appointed from the United States. Among those delegates was Lucretia Mott, a leading female abolitionist from Pennsylvania and a staunch advocate of women's rights in America. Even though in the end women were not allowed to take an active part, but only to assist from the balcony of the hall where the convention was held, the controversy did much to stir the public opinion over the link between the fight against slavery and the fight for women's rights. It was also stirred as a result of Garrison's decision to join, together with his delegation, the women in the balcony, as a sign of protest.[63]

In the following days, the debates led to important decisions. Notable, thanks to the activism of American abolitionist and churchman James G. Birney, were the condemnation of slavery as "a sin against God" and the resolution to use the considerable influence of the English Church on the American churches against slavery in the United States.[64]

Clearly, since British emancipation had been achieved, American slavery had become the priority in the activities of Atlantic abolitionism. To this end, together with involving the religious authorities to mobilize American public opinion, the convention also promoted the encouragement of the use of free labor in the production of cotton and sugar in the American plantations. On the other hand, an American delegate formally asked Daniel O'Connell to address Irish Americans, some of whom were "the principal supporters of slavery," with the purpose of converting them to the abolitionist cause.[65]

Along with Garrison and O'Connell, Ashurst was also present at the World's Anti-Slavery Convention, though he participated in only a few debates. Mazzini did not attend the convention, but was aware of it and wholeheartedly supported it.

In practice, the World's Anti-Slavery Convention in London in 1840 signaled the peak of Atlantic abolitionism. It allowed different Atlantic progressives to actually work together toward the same objective—the abolition of slavery—while making them increasingly aware of the Atlantic dimension of other equally important issues, such as women's rights.

The repercussions on the public opinion were enormous, and the success of the convention led to a second one in London in 1843. However, the main issue that Atlantic abolitionism addressed—slavery in the United States—continued to exist, with alternate fortunes, for the following two decades.

As the political conflict between the U.S. North and South exacerbated in the following years, Garrisonian nonresistant abolitionism suffered a major setback, either because it was not radical enough in its actions against slavery or because it was too radical in its actual stand against slavery. On one hand, especially in the 1850s, radical abolitionists such as John Brown, tired of the repeated failures in efforts to move the American public opinion, engaged in direct violent action against slaveholders. On the other hand, in the 1840s and 1850s, a series of moderate antislavery coalitions became increasingly the political expression of a majority of voters in the North. The last of this coalition, the Republican Party, saw the rise to prominence of Abraham Lincoln and then his election to President of the United States, an event that prompted the secession of the majority of the southern states and led to the Civil War. In the course of the war, Lincoln signed the Emancipation Proclamation of January 1, 1863—an act that fulfilled Garrison's and the abolitionists' prophecies on the end of American slavery—leading to the liberation of 4 million slaves.[66]

The 1848 European Revolutions

If the World's Anti-Slavery Convention signaled the high tide of Atlantic abolitionism in 1840, it was the European-wide revolution of 1848—the Springtime of the People—that signaled the high tide of European nationalism. Even though scholars have rarely connected the two movements, the democratic demands that national revolutionaries voiced in different countries in Europe in 1848 through 1849, primarily the recognition of oppressed nationalities, were very much in line with what Atlantic progressives who were actively involved in radical antislavery thought as the indispensable requirements for humankind's march toward progress.

According to C. A. Bayly, while "political and nationalist demands ... were at the forefront of the intentions of the educated participants [to the national movements] ... the leaders of 1848 demanded suffrage reform and self-determination," and thus "the language of the rights of man and the invocation of the revolutionary tradition" were very much at the center of political action. This language had its distant roots in the principles set out during the Age of Revolutions on both sides of the Atlantic, even though, in the European case, the specific reference to the French Revolution and its legacy was much more direct. At the same time, there is little doubt that the actual language that European revolutionaries used to voice their demands for national self-determination and suffrage reform was a language inherited directly from the struggle against slavery and from Atlantic abolitionism.[67]

Thus, the expressions that characterized the discourses on national freedom were borrowed from the speeches and pamphlets that talked about human freedom. But it was much more than simply a linguistic issue, since for Atlantic progressives such as Garrison, Ashurst, O'Connell, and Mazzini, the link between slavery's abolition and national self-determination was almost self-evident. For this reason, they were actively involved on more fronts in different progressive causes, and especially in the radical antislavery cause, through their participation in Atlantic abolitionism, while supporting also European democratic nationalism.

When the 1848 European revolutions began, Garrison and Ashurst sustained enthusiastically Europe's national movements from the columns of their papers or in their speeches, while Mazzini became a protagonist in the nationalist struggles in the continent. O'Connell had died shortly before then, in 1847, in time to see Ireland strangled by the Great Famine and England taken over by the Chartist Movement.

In practice, in rapid succession, between January and October 1848, movements for national self-determination and suffrage reform broke out in Sicily, Naples, Paris, Frankfurt, Berlin, Paris, Vienna, Prague, Budapest, Venice, and

Rome, enveloping the entire continent into a massive revolutionary wave. The link between national self-determination, democratic government, and the abolition of slavery became even clearer, and was sealed once and for all, when the new democratic government that established the French Republic in February 1848 decreed the emancipation of all the slaves in the French colonies.[68]

The struggle for national self-determination in Italy, where Mazzini hoped to establish a democratic republic, prompted particularly sympathetic reactions from all the Atlantic progressives. Mazzini himself was actively involved firsthand in the revolution that led to the formation of the Roman Republic and guided the republic's government for a few months between March and July 1849. He led the movement to issue a democratic constitution that reversed centuries of papal abuse, abolished the Inquisition and ecclesiastical privilege, and protected freedom of thought and religion.[69]

Even though the Roman Republic ended tragically, Mazzini earned the admiration of both his fellow European revolutionaries and the American radicals. Indeed, it was at this point that Garrison elected any occasion to magnify Mazzini's achievements by publishing the carefully documented articles written by Ashurst. However, after 1849, Mazzini became more and more isolated, as fewer and fewer Italian patriots were convinced of the possibility—some even of the necessity—of building a democratic republic. Thus, throughout the 1850s and 1860s, a moderate solution to the Italian national problem—with the House of Savoy at the head of a unified and free Italian monarchy—became increasingly popular. Eventually, it became reality when Victor Emmanuel II proclaimed the Kingdom of Italy in March 1861, while Mazzini, keeping his republican convictions, wound up in exile again.[70]

By 1865, when the American Civil War concluded with the emancipation of all the slaves in the southern part of the United States, while Italy had become a free nation only four years earlier, only the youngest two of the four Atlantic progressives were still alive: William Lloyd Garrison and Giuseppe Mazzini. If, in the case of the United States, we can say that Garrison obtained his ultimate victory with slave emancipation, the same was not true for Mazzini, who saw his dream of a democratic republic in Italy—let alone that of a communion of democratic republics in Europe—pushed aside by the emergence of an Italian monarchy. Still, in the 1860s, the formation of a liberal Italian nation, almost contemporary with the reunification of a wholly free United States, concluded symbolically a long era of democratic struggles that had started with the formation of the new abolitionist movements in the late 1820s and early 1830s.

By the 1860s, slave emancipation and national self-determination had become part of a distinctively western and Atlantic tradition of freedom. 
However, in many ways, it was a less radical tradition than the one Atlantic progressives—first through the network of Atlantic abolitionism, and then through the support of European democratic nationalism—had worked hard to create and maintain a few decades earlier.

Atlantic Progressives and Their Entangled Histories

The analysis of the lives of the four Atlantic progressives we have examined shows how much it is possible to gain from an interrelated analysis of American, British, and Irish abolitionism and Italian and European nationalism in a broader, comparative perspective. The lives of Atlantic progressives are almost a paradigmatic case of a type of historical method called *histoire croisée*, for "entangled history," because they were connected to one another through networks that spread not just across the Atlantic, but also across Europe. The Atlantic progressives' networks built on established connections that dated back to the Age of Revolutions, and their fundamental beliefs had been shaped by a commitment to those ideals expressed first in the American and French Revolutions.[71]

But, from the 1820s onward, Atlantic progressives gathered around a new view of the struggle against slavery. By supporting an immediatist, radical version of abolitionism, they discovered that radical activists—not just abolitionists, but also democratic nationalists—shared some of their most profound beliefs in both the United States and Europe.

While Atlantic progressives such as Garrison were at the forefront of radical activities aimed at the abolition of slavery in the United States, Atlantic progressives such as Ashurst were at the forefront of parliamentary reform and involved in a variety of reform movements—from abolitionism to women's rights—in Britain. At the same time, Atlantic progressives such as O'Connell were at the forefront of movements aimed, through parliamentary reform, at ameliorating the conditions of the Catholic Irish population. Still other Atlantic progressives, such as Mazzini, fought their democratic nationalist battle to create an Italian republic.

What united all these efforts was a common unwavering belief in progress, and therefore in the ideal that all the causes that led to humankind's gain in terms of human rights were self-justified struggles, for they allowed more people to share in the benefits that progress inevitably brought. It was a highly teleological view, but also one that succeeded in creating the preconditions for the creation of large, international and transatlantic movements for freedom, as Atlantic abolitionism and European democratic nationalism effectively were.

These movements reached their peak in the 1840s, first with the World's Anti-Slavery Convention in London in 1840, and then with the European-wide revolutionary upheavals of 1848 and 1849. In both cases, the networks created by Atlantic progressives showed their strength through their direct participation or through their firm support for momentous events in the name of common ideals. Only by looking at these premises is it possible to understand how, later on, in the 1860s, slave emancipation in the United States and the creation of a liberal Italian nation in Europe were events that had a truly Atlantic dimension.

Chapter Six

The American Civil War, Slavery, and Emancipation in a Nation-Building Age

Increasingly, scholars are adopting a transnational view in studying the American Civil War. Recent syntheses of American history in transnational perspective—such as the ones written by Thomas Bender, Carl Guarneri, and Ian Tyrrell—have placed the Civil War in the context of contemporary events in Europe.[1]

The central decades of the nineteenth century, in the period between 1848 and 1870, saw the unfolding of the classic age of nation-building in Europe. The rise of nationalist movements throughout the continent during the revolutionary period of 1848 to 1849 was followed by the consolidation of two new large nation-states in Italy and Germany. The significant connection between the American Civil War and the latter two events was well understood by Michael Geyer and Charles Bright, who, as early as 1996, had written that "the nation-making, state-creating, industrial and expansive outcome of war ... makes the middle of the century wars of central Europe and North America similar" and comparable. Essentially, this is a similar point to the one that transnational historians would argue in their treatment of the American Civil War as a process akin to Italian and German national unifications. As Carl Guarneri has written as recently as 2007, "the Northerners' plan to unite the states under a strong central government ... aligned the American struggle with wars of national unification that were occurring in Europe and the Americas in the same decades."[2]

Yet, we cannot forget that, despite all these striking similarities, the main difference between the American and European situations is that the Civil War did not merely reunify a nation, but it also brought about the emancipation of 4 million slaves. Indeed, U.S. slave emancipation was such a significant event on a global scale that a number of scholars have focused, for at least the past 20 years, on specific comparisons between the uniquely violent process of emancipation during the American Civil War and the relatively peaceful ending of slavery or serfdom in other areas of the Americas or in Europe.

Currently, there is an ever-increasing scholarly interest in the comparative history of American emancipation, proceeding from Eric Foner's 1983 seminal study *Nothing but Freedom.* Subsequently, later studies by scholars such as Steven Hahn, Peter Kolchin, Rebecca Scott, and Stanley Engerman have added a great deal to our understanding of how the ending of slavery in the United States should be seen in the wider context of events and measures occurring in the "age of emancipation." Like the age of nation-building, this also occurred during the central decades of the nineteenth century.[3]

In this connection, it is particularly important to recognize that the age of emancipation and the age of nation-building in the Americas occurred as a result of transformations traceable to the making of the second slavery. At the end of the eighteenth century, the 1791–1804 revolution in St. Domingue and the consequent emancipation of Haiti signaled the end of the colonial system of slave production in the New World. Together with the abolition of the British and American Atlantic slave trades in 1807 and 1808, this led to the beginning of the nationalization of slavery as a major factor of national economic expansion, particularly in the U.S. South with cotton and sugar, in Cuba with sugar, and in Brazil with coffee.[4]

As discussed in the previous chapters, due to their command of the world economy through their respective agricultural products and to the particular technologically driven and capitalist-oriented nature of the national slave-based economies present within them, the nineteenth-century U.S. South, Cuba, and Brazil were at the heart of the phenomenon that Dale Tomich and Michael Zeuske call *second slavery.* It is important, though, to recognize that "the world-historical processes that transformed the Atlantic World between the 1780s and 1888" not only resulted in the "formation of highly productive new zones of slave commodity production" in the U.S. South, Cuba, and Brazil, but also eventually led to their demise through processes of emancipation that happened in different ways and at different times. Essentially, the three regions at the heart of the second slavery went first through comparable, and somewhat similar, processes of nationalization of slavery, and then through comparable, though somewhat different, processes of nationalization of freedom.[5]

Within this context, we must think of the American Civil War as a process that, through the creation of a stronger, centralized, and reunified nation-state, led to the making of slave emancipation, and thus to the victory of the nationalization of freedom over the nationalization of slavery. On one hand, the Republican Party was very instrumental in this process with its mild policy of prevention of nationalization of American slavery through its containment within the South. On the other hand, Lincoln was certainly the one who gave the final blow to the Confederate States of America's ultimate project of nationalization of slavery through the 1863 Emancipation Proclamation.[6]

During the Civil War, the southern project of nationalization of slavery achieved only temporary fulfillment. First, in 1860 through 1861, there was the transformation of the regional and sectional southern nationalism into a Confederate slaveholding nationalism with the secession of the majority of the southern states from the Union. Then, in 1861, came the creation of a fraught and internally divided Confederate slaveholding nation separate from the United States. Four years later, as a result of the 1865 Union victory in the Civil War, the Republican project of nationalization of freedom achieved the status of permanency within the United States, and changed forever the very idea of the American nation, truly unifying it for the first time.[7]

Thus, the process of nation-building and the process of emancipation as they occurred in the United States through the American Civil War are very much related to one another, and they should be studied and understood in conjunction. Yet, despite the fact that the two scholarships—one focusing on the Civil War in an age of nation-building, and the other on U.S. slave emancipation in an age of emancipation—aim at building the foundations of a comparative perspective of two aspects of the same event—the American Civil War—contact between scholars working on these two fields has been scant. Also, studies claiming the comparability of one of these two aspects have made generally little or no reference to the comparability of the other aspect as well.

In this chapter, I will suggest possible ways to combine these two approaches. I argue that, in order to build the foundations of a thorough comparative history of the American Civil War, we need to take into account, in equal measure, its characteristics as a nation-building process and the significance of the central event of emancipation at its core, in comparative perspective. In sum, I believe that only by directing our investigative efforts toward explaining the multidimensional nature of the American Civil War as a conflict with a clearly global significance can we begin to understand the best and most accurate way to proceed to write its comparative history (see Map 5).

In a 1987 article entitled "Thesis, Antithesis, Synthesis: The South, the North, and the Nation," Carl Degler used Hegelian dialectic to argue that he saw "American history as the interaction of North and South—a Hegelian synthesis that has emerged from the dialectic of a southern thesis and a northern antithesis." Taking inspiration from that article, I wish to use a similar Hegelian dialectic to analyze through a comparative perspective the significance of the American Civil War in the nineteenth-century Euro-American world.[8]

In particular, I use historical comparisons in order to elucidate the significance of the following three assumptions:

Map 5 The Age of National Unifications and Emancipations in Europe and the Americas (1860–1890)

Map 5a Europe

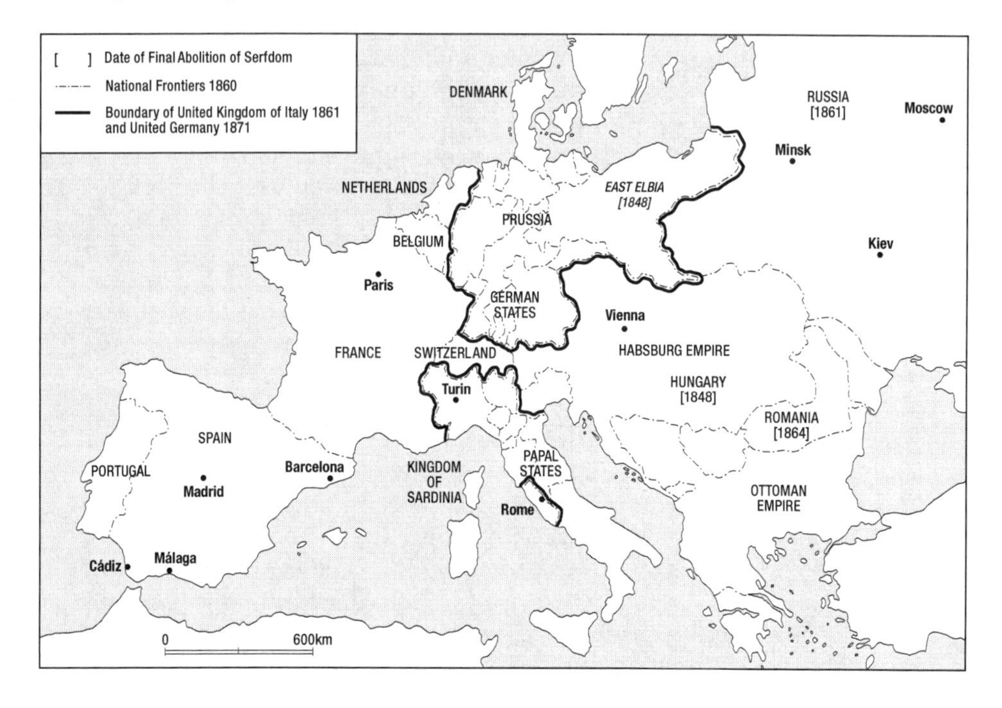

Thesis: The ideology of the Republican Party in the United States until the American Civil War had some characteristics in common with liberal nationalist ideologies in mid-nineteenth-century Europe and liberal ideologies in Latin America, particularly its moderate and non-revolutionary character.

Antithesis: The occurrence of slave emancipation during the American Civil War was a revolutionary event with radical features that set it apart from most other processes of emancipation from unfree labor occurring in the nineteenth-century Euro-American world. It was in clear opposition with the mild antislavery policy of the antebellum Republican Party.

Synthesis: The process of nation-building in the United States, through the reconstitution of the Union during the Civil War—in which Abraham Lincoln played a crucial role—is most unique compared with similar and contemporary processes of nation-building. This allowed the transformation of the mild antislavery Republican Party ideology into one that had more room for Radical Republicans, and thus allowed the possibility of a revolutionary act such as slave emancipation.

Map 5b The Americas: The Civil War United States

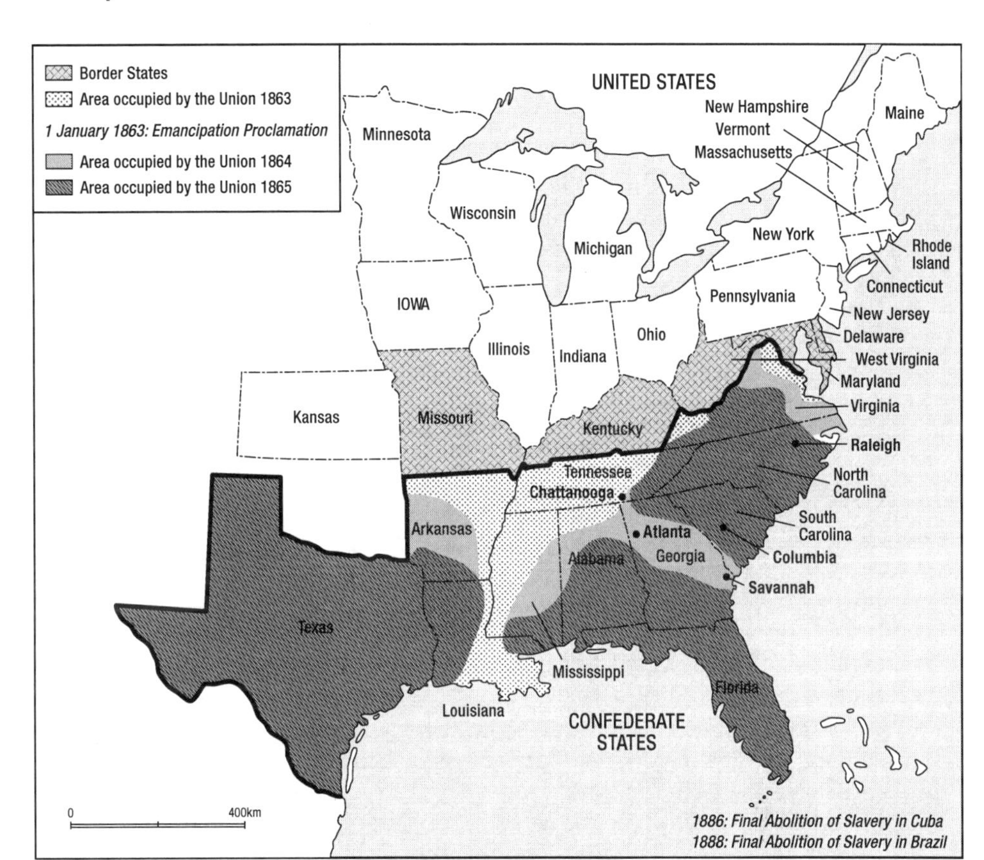

THE U.S. REPUBLICAN PARTY AND EURO-AMERICAN LIBERAL NATIONALISM

In 2011, in his recent textbook on the history of the United States, *Give Me Liberty!,* Eric Foner wrote that "the American Civil War was part of a broader nineteenth-century process of nation building. Throughout the world, powerful, centralized nation-states developed in old countries ... and new nations emerged where none had previously existed." However, still according to Foner, what distinguished the American nation that took its definitive shape under Abraham Lincoln's guidance during the Civil War was the fact that this particular process of nation-building was not based on the idea of freeing or unifying a particular people or ethnicity, but rather on a "set of universal ideas, centered on political democracy and human liberty."[9]

We can take Foner's words as guidelines for making some reflections on both the past and the current scholarship on the American Civil War era in comparative perspective. In fact, it was precisely by focusing on the universality of the idea of human liberty in nation-building struggles that David Potter first argued in 1968 about the comparability of the Civil War era in the United States with nineteenth-century European nationalist movements. However, while many of the points Potter made more than 40 years ago are usually included in short comparative treatments by current transnational and comparative historians, Potter's argument on the specific comparability between the processes of nation-building in the United States and Europe in the early 1860s is still only rarely given the importance it deserves. On the other hand, given Potter's emphasis on the liberal aspects of nineteenth-century nationalism, a thorough comparative view of the American Civil War should make reference not just to European, but also to Latin American varieties of liberal and liberal nationalist ideologies as they expressed themselves in the first six decades of the nineteenth century.[10]

Thus, when looking at the Civil War era in the United States in comparative perspective with both Europe and Latin America, a particularly important question is whether the principles supported by the Republican Party, which would guide the Union and reconstitute the American nation, had common traits with the principles of the movements supporting liberal nationalism that sprang up all over the Euro-American world, starting from the early decades of the nineteenth century.

The Republican Party in Antebellum America

The Republican Party rose to the national scene in the United States in 1854. Antislavery members of the Whig Party, together with Democrats and members of the Free Soil Party, gathered together to form an antislavery coalition and decided to call themselves Republicans, in honor of Thomas Jefferson, the ideological father of the American Republic. Effectively, the Republican Party was the first major political party to publicly condemn slavery as a "relic of barbarism," in the words used in the 1856 election campaign, and to present a program with which its members intended to oppose slavery's expansion.[11]

As Don Fehrenbacher has noted in *The Slaveholding Republic* (2001), the attitude of the Republican Party toward slavery constituted a real rupture in the antebellum American political tradition. While the Constitution contained only an ambiguous approval of the slave system in the South, the attitude of the U.S. government, whose key posts were controlled by the planter elite, and of the major political parties on the problem of slavery's expansion had been, for the most part, favorable to southern interests.[12]

The Republicans were the first representatives of a major political party to oppose slavery in a moral sense, to openly criticize the influence of the slave power on the federal government, and to attempt to put an end to this influence politically. It is also true, however, that the Republican Party's antislavery attitude was not as radical as that of the abolitionists, since their program did not contemplate the abolition—let alone the immediate abolition—of slavery. Therefore, in this sense, Republicans were hardly revolutionaries. Rather, we should see them as moderate, akin to the way most liberals in the Euro-American world would have thought of themselves, since they also did not support political programs with sudden radical changes at their core.[13]

The ideology at the heart of the Republican Party's political program was particularly complex and included different elements, which had originated from the various types of opposition to slavery that characterized American politics since 1848. At its heart were the concepts expressed in the motto "Free Soil, Free Labor, Free Men," which was the official motto of the Republican Party. In his 1970 book by the same title, Eric Foner magisterially analyzed these concepts, arguing that the Republican Party's antislavery ideology was based on the idea of moral and material superiority of the free economic and labor system that characterized the North. The North was essentially an industrializing capitalist system, in which everyone presumably had equal opportunities of social advancement.[14]

Following Foner, we can distinguish in the Republican Party's ideology three different concepts that were collapsed together in order to support a particular and novel form of antislavery thought:

- The concept of *free soil,* which dated to the 1848 Free Soil Party and referred to the objective of keeping slavery away from the new western territories
- The concept of *free labor,* which indicated the superiority of the northern economic and social system
- The concept of *free men,* which indicated the moral superiority of a system in which everyone was considered equal, instead of being divided between free and slaves, and whose very distant origins were in the abolitionists' intransigent attitude toward the inequalities caused by slavery[15]

In essence, these three elements combined within the Republican Party's antislavery ideology of "Republican nationalism," as Peter Parish has called it. The Republicans constituted a novelty in supporting a political formation aspiring to become not just a national party, but the very party that would lead the American nation and would refashion it according to these principles.[16]

Despite these crucial elements of novelty, though, it was only with Abraham Lincoln's rise in the late 1850s that the Republican Party completed the process of systematizing its ideology and of popularizing its particular antislavery thought, consolidating it in a distinctive form of Republican nationalism. According to Parish, "the key to Republican nationalism may be found in the notion of the Republican Party as a party of improvement ... [since] Republicans espoused the cause of human betterment and, in the process, linked the idea of the Union to the cause of such betterment."[17]

Similar to European and Latin American liberals, American Republicans, akin to their forerunners the Whigs, put the accent on progress, both in material, or economic, terms, and in moral terms. Also, comparably to European and Latin American liberal and liberal nationalists, American Republicans linked this progress—which included the essential feature of opposition to slavery—to the very idea of the nation, or, in the case of the United States, the Union.[18]

Encompassing within itself the different elements of republican ideology and referring to contemporary ideas that informed liberal nations and liberal nationalist struggles throughout the Euro-American world, Lincoln's focus was firmly on the indissolubility of the Union, or the American nation, as a thought, an ideal, and a concrete political institution.

Lincoln best expressed his view in Springfield, Illinois, in 1858, in the "House Divided" speech, which was his acceptance speech for the Republican candidacy to the Illinois post in the Senate. In that speech, Lincoln expressed in definitive form the main guidelines of his thought, which then became the official doctrine of the Republican Party. On that occasion, Lincoln argued forcefully against slavery's expansion in the western territories. Referring to a New Testament analogy, he compared the Union to a "house divided within itself" and whose situation should find a solution either in favor of liberty, as he hoped, or in favor of slavery, as he made clear through the words "this government cannot last forever half *slave* and half *free*." With this inspiring analogy, Lincoln placed at the center of what Ian Tyrrell has termed his "politics of moderate, liberal nationalism" the integrity of the Union, and focused specifically on the consequences of the problem of slavery's expansion (rather than on abolition of slavery) as a threat to both the integrity and commitment to freedom of the American nation.[19]

Liberal Nationalism in Nineteenth-Century Europe

There is more than one parallel between Lincoln's moderate vision of the reunified American nation as one founded on liberty and the principles of the Declaration of Independence and the contemporaneous European liberal nationalist movements. For a start, the American Declaration of Independence itself was, together with the 1789 French Declaration of the Rights of Man and Citizen, the very foundational document of the doctrine of liberal nationalism and a constant source of inspiration for Europe's liberal nationalists.

Given the emphasis of liberalism on the individual and of nationalism on the community, at first sight, the two ideologies might have proven incompatible. However, in nineteenth-century Europe, in the words of Mark Hewitson, "most liberals viewed the nation-state as a natural entity." This was to the point that even the never-ending debate over whether their nationalism was civic or ethnic seems to pale and lose all significance.[20]

Like the American Revolution, the French Revolution in its early phase established that a nation was a voluntary community of responsible citizens and that it was to protect the basic individual and universal rights. Together with the important influence of British constitutional thought, these basic ideas continued to be at the center of liberal nationalist ideology for the best part of seven decades in nineteenth-century Europe. This ideology inspired struggles against autocratic rule by demanding the formation of nations based on "freedom of the press, the rule of law, constitutional governments, and greater political participation in the affairs of the state."[21]

The nationalist struggles based on these demands reached their peak with the European revolutions of 1848 through 1849. However, these revolutions also showed that, even though they were theoretically supporting universal human rights, and they were temporarily able to join more radical democratic or socialist movements in order to overthrow autocratic rule, liberal nationalists throughout Europe were not prepared to make concessions to the wider masses in terms of sharing the actual governmental power. As a consequence, post-1848 European "liberal national movements," in the words of Stefan Berger, "practiced a range of exclusionary mechanisms directed against the lower social classes, which were perceived as lacking the preconditions for political activity, namely education and property," thus basically restricting as much as possible suffrage and access to vote. In this way, Europe's liberal nationalists showed that their ideology was antirevolutionary, antiradical, and moderate, in the sense of privileging the rights of the "men of property and standing" over those of the dispossessed multitudes, whom they were prepared to help only through paternalistic programs and charitable associations.[22]

Though in a different context and with different social and cultural references, the moderate type of nationalist ideology that post-1848 European moderate liberals embodied had more than one point in common with Lincoln's moderate Union-focused nationalism in the 1850s. In both cases, the emphasis was on a clear dichotomy between the theoretical radicalism of recognized general principles—on one hand, universal rights, and on the other hand, human freedom—and the moderate dimension of the actual political practice—on one hand, the actual restriction of suffrage, and on the other hand, a general opposition to slavery's expansion, rather than a demand for its abolition.

A typical case of a European liberal nationalist whose ideology was moderate in this sense is that of the prime minister of the Kingdom of Sardinia, and future Italy's first prime minister, Count Camillo Cavour. Unlike Lincoln, Cavour and his political allies—significantly named Moderate Liberals—faced the problem that Italy in 1859 was not just divided politically and ideologically, but was also under the strict control of the Habsburg Empire. Thus, the actual process of national unification involved a war for liberation against an enemy that was much more powerful than the relatively small Kingdom of Sardinia.[23]

At this time, Cavour's idea of a unified Italy corresponded largely with the formation of an Italian state through the expansion and spread of Piedmontese liberal institutions. By this, he meant fundamentally a constitutional government with members of parliament elected on the basis of substantial property requirements, which would eventually lead to the formation of a Kingdom of Italy. In other words, faithful to his moderate positions, Cavour sought not to force the completion of the process of nation-building if the circumstances did not require it. This was akin to Lincoln's moderate attitude of nonintervention in regard to slavery at the outset of the American Civil War.[24]

By the same token, even though advanced by nineteenth-century continental European standards, the concepts of civic liberty and political participation that Cavour and the Moderate Liberals shared were as far as they could be from the idea of a republican nation based on universal suffrage that characterized their political opponents, the Democrats. Therefore, in this sense, their moderate positions were more than tinged with strong elements of conservatism.

Liberalism in Nineteenth-Century Latin America

Akin to, and contemporaneous with, the evolution of liberalist nationalist thought and practice in Europe, liberalism in Latin America went through different stages. In a similar fashion, this process led to an ultimate conservative involution in most of the countries that had freed themselves from Europe's colonial yoke by the time politics entered the central decades of the nineteenth century.

After 1825 and the achievement of independence, with the beginning of the early national period (1825 through 1850), the creole elites of landowners and intellectuals that ruled the newly born Latin American republics found themselves with the difficult task of governing countries that were extremely large and diverse. Additionally, the majority of the population was illiterate and therefore unable, in the mind of the political classes, to contribute meaningfully to the life of the nation.[25]

Still, initially, the liberalism that characterized the elites in power led them to make repeated attempts to create constitutional frameworks in order to form the bases of representative electoral systems. This was despite the pressures to turn to authoritarian rule, brought upon them by a number of military strong men (*caudillos*). To this end, in the constitutions drafted in different countries, initially the Latin American elites supported, consistently with the aims of liberal elites throughout the Euro-American world in the first half of the nineteenth century, "enlightened goals which generally favored elections, equality before the law, freedom of speech, educational reform, and industrialization," in the words of Will Fowler. At the same time, suffrage was theoretically extended to the majority of the male population.[26]

However, after 1850, and also consistently with the general trend of liberal thought in the Euro-American world, as a result of the recrudescence of social division between different classes, the rapid rise of *caudillos* as the guarantors of order and stability, and the continuous rift between supporters of stronger centralized governments and supporters of devolution of power, the elites in power in the different Latin American countries turned to a more conservative form of liberalism. In practice, the central decades of the nineteenth century saw the resolution of the ongoing struggle between liberals and conservatives, with the triumph of the liberals, but also, at the same time, with the incorporation of a number of conservative features in Latin American liberalism.[27]

As a result, the liberal elites that came to power in different Latin American countries, especially from the 1870s, were keen to stress the need for "order and progress." This referred to the need for a stable political system that favored the economic improvement of the country through support to the activity of entrepreneurial landowners, industrialists, and bankers. In this new type of political system, universal suffrage was no longer an option, and, significantly, the constitutions approved during this period contained provisions for a suffrage restricted effectively to the propertied and educated oligarchies that ruled the different Latin American countries.[28]

Comparison between the ideology of the U.S. Republican Party and both European liberal nationalism and Latin American liberalism highlights the similarity between particular features of what Peter Parish has called "Republican nationalism" and liberal nationalist thought in the Euro-American world. Specifically, by the central decades of the nineteenth century, liberal nationalism in Europe and liberalism in Latin America had evolved into conservative, or moderate, versions of the ideologies that had inspired the European revolutions of 1848 to 1849 and the struggles for Latin American independence in the 1820s.

By the time both continents had reached the central decades of the nineteenth century, faced with the prospect of a class war prompted by radical programs, the European and Latin American middle classes had closed ranks and focused on a narrow interpretation of the idea of protection of liberties and political participation, effectively excluding the majority of the population. At the same time, though, in Europe and Latin America, liberal intellectuals and politicians continued to agitate for constitutional reform and parliamentary representation, while sharing a deep belief in the link between economic and civil progress and nation-building.[29]

Keeping in mind the differences due to the existence of universal male suffrage in the United States and its absence in Europe and Latin America at this time, we can say that these points can be fruitfully compared with the ideology of Republican nationalism advanced by Lincoln and the Republican Party in the United States. Of particular interest is the link between economic and civil progress and the cause of the Union, or the American nation, which was clearly a top priority in the Republican agenda. At the same time, we can also link the more conservative, or moderate, features of European liberal nationalism and Latin American liberalism to the ideology of the Republican Party if we think that the party was committed to a general opposition to and containment of slavery, not to its abolition. This effectively excluded a large section of the American population from the enjoyment of basic liberties and civil rights, comparably to the way European and Latin American liberals excluded large sections of the populations of different countries on the basis of property rights and literacy.

EMANCIPATIONS: ENDING SLAVERY AND SERFDOM

An immensely important outcome of the American Civil War was the 1863 Emancipation Proclamation, which eventually resulted in the freedom of 4 million slaves. Without minimizing the importance of this event, we should see it in its proper context. It occurred at a time—the second half of the nineteenth century—when emancipation from unfree labor extended progressively to all the countries where slavery or serfdom still existed.

Building on the important foundations established by the Atlantic abolitionist movements of the earlier part of the century, and by previous momentous pieces of legislation (particularly the 1833 British Act of Emancipation and the 1848 law that freed all the slaves in the French colonies), movements that opposed unfree labor gathered new momentum, but not until the later part of the period of 1850 to 1890.

In the early decades of this period—the 1850s and 1860s—Atlantic abolitionism was less effective, and the collaboration and links between abolitionists across the ocean never again reached the levels of 1840, the year of the first World's Anti-Slavery Convention in London. Still, through their established networks, abolitionists on both side of the Atlantic were able to focus their attention on the rapid and dramatic changes in the internal situation of the United States. These changes, as a consequence of the annexation of the slaveholding republic of Texas and the subsequent Mexican War (1846 through 1848), led to the centrality of slavery, and specifically the issue of its expansion, in the 1850s political conflict between the U.S. northern and southern states.[30]

At the same time, also thanks to the influence of abolitionists in the British Parliament, the 1840s witnessed an increase in British diplomatic efforts to end the slave trade, to the point that its abolition became part of the British government's official foreign policy. Between 1835 and 1850, several European and Latin American countries signed a "cascade of new treaties," in Seymour Drescher's words, aimed at patrolling against the slave trade, largely as a result of British pressures. This shows the truly global dimension of the British abolitionist project. Also, in a way, it is part of the story of Atlantic abolitionism.[31]

The British government put particular pressure on Brazil, whose flourishing Atlantic slave trade saw the active participation of a number of U.S. citizens. Effectively, the British Navy played a major role in the Brazilian slave trade's eventual abolition in 1850. Despite this, British abolitionists were internally divided, particularly over the need for high protective duties to keep up colonial sugar production in the British West Indies, where it fell consistently throughout the 1840s. In truth, the leaders of the BFASS were all in favor of protectionism, not wanting to admit the effective failure of "the mighty experiment." Thus, also as a result of this, according to Seymour Drescher, "the early 1840s ... marked a recession in the fortunes of British antislavery all around the Atlantic basin," which was a trend that would continue throughout the 1840s and 1850s.[32]

Still, in the two decades preceding 1860, both British and French warships patrolled the coasts of West Africa, as a result of 1831 and 1835 agreements the two powers had signed on the suppression of the slave trade. By 1845, the United States and Portugal had also joined in what became, effectively, an international antislavery task force. Yet, only four years later, France, now under Napoleon III, resumed covertly its own slave trade.[33]

Throughout the 1840s and the 1850s, the British Navy remained the dominant power in charge of suppressing the slave trade in West Africa. Britain succeeded in destroying a number of trading stations and forts, and eventually also in signing several treaties leading to suppression of the slave trade with African rulers. British pressure also led the Ottoman sultan to suppress the trade in 1857.[34]

In Latin America, by 1850, among all the Spanish-American countries, only Mexico, the Central American federation, and Chile had abolished slavery. In particular, Mexico had already abolished the slave trade in 1824 and slavery in 1829. Therefore, the creation of a slaveholding Texan region filled with U.S. immigrants in the northern part of Mexico's territory in the early 1830s was an act in open defiance of Mexican law—one that eventually led to Texas's independence and the Mexican War, with the country's 1848 loss of 525,000 square miles to the United States. Then, in the ten years between 1851 and 1861, all the remaining Spanish-American republics that had not abolished slavery at the moment of their formation in the 1820s—aside from Paraguay, which waited until 1869—also proceeded to free their slaves.[35]

By the time the American Civil War started, slavery had disappeared from most of the Americas. It remained entrenched only in Cuba and Brazil, where it had thrived under the second slavery, and where it was finally abolished in 1886 and 1888, respectively.

Concurrent with this gathering of momentum in the movement for the abolition of slavery was the parallel and contemporaneous movement for the abolition of serfdom, which built on the example of provisions made at the time of the 1848 revolutions to free the peasantry in Prussia and the Habsburg lands. As a result, by 1860, serfdom still existed only in Romania

and Russia. In the space of five years, serfdom would disappear from these areas as well—first in Russia in 1861 and then in Romania in 1864. Despite the fact that serfdom by this time was restricted to these two countries, the emancipation of eastern European serfs affected a much larger number of individuals—certainly, more than double—than slave emancipation in the Americas. The number of serfs emancipated in the 1860s was well over 20 million. In contrast, the number of slaves emancipated in the period of 1850 through 1890 was much less than 10 million.[36]

The Making of Emancipation in the Civil War United States

In the United States, in the course of the four years of the American Civil War (1861–1865), a combination of factors led Lincoln to support the idea of slave emancipation. This was an unprecedented development, and one contrary to the previous mild antislavery policy of the Republican Party.

First of all, the slaves themselves, who had fled the southern plantations in the Confederate nation in large numbers from the beginning of the war and had reached the Union camps, posed a problem regarding their legal status for Union officers. In 1861, General Benjamin Butler coined the definition "contrabands of war" for the runaway slaves. The same year, Congress passed the First Confiscation Act, according to which all rebel property, including slaves, was liable to be seized. In 1862, a Second Confiscation Act stated that all slaves of Confederate masters were to be considered free and called for the seizure of Confederate property.[37]

Therefore, the increasing pressure exercised by the growing problem of runaway slaves turning up in Union camps became one of the main factors that created the necessary conditions for the Union government's and Lincoln's gradual leaning toward a more radical position regarding the problem of slavery. At the same time, the Republican Party itself was increasingly radicalizing its views, also under the continuous pressure exercised in Congress by Radical Republicans, many of whom were former abolitionists, in order to reach quickly a solution aimed at the complete end of slavery.[38]

Lincoln's view also changed from a position privileging the safeguarding of the Union at all costs, including leaving slavery intact, to one in which he contemplated the idea of issuing an Emancipation Proclamation, especially since he saw that the latter could bring a long and costly war to a quicker end. Thus, on September 22, after the Union victory at the battle of Antietam, Maryland, Lincoln released a first version of the Emancipation Proclamation, which he then modified and signed in its definitive form on January 1, 1863.[39]

As many scholars have noted, Lincoln's Emancipation Proclamation was, first and foremost, a war measure that abolished slavery only in those Confederate territories not yet reached by the Union Army, and therefore, could not contribute to its enactment. It did not mention the slaves who were already in Union-occupied territories. Still, there is no doubt that it was a revolutionary act of immense significance. One reason was that, with an unprecedented action, it abolished legally slavery, both immediately and in perpetuity. Also, as a consequence of the Emancipation Proclamation, wherever they arrived, the Union soldiers now acted as troops of a liberation army, officially instructed by the Union government "to recognize and maintain the freedom of such persons [the slaves]," as the Proclamation stated.[40]

The effects of the Emancipation Proclamation on the South were devastating. The slave economy and society, already in difficulties as a result of the Confederate nation's inability to face a modern, industrial, conflict such as the American Civil War, quickly collapsed. With the increasing number of southern men drafted to the front, fewer and fewer slaveholders were able to mind their properties, which were left in the hands of plantation mistresses and overseers. Whenever they could, slaves took advantage of the situation, by either openly rebelling or simply running away. Wherever the rumor arrived that Lincoln had released the Emancipation Proclamation, slave rebellions and escapes became more frequent. This was especially the case in those areas that were closer to the Union lines in the states of the upper South.[41]

In time, the continuous disruption of the plantation economy caused by the slaves' activities gave a decisive contribution to accelerating the end of the slave system, and therefore the support for the Confederate States, whose final defeat occurred in April 1865. In December 1865, Congress ratified the Thirteenth Amendment, which prohibited slavery throughout the United States.[42]

The End of Slavery in Cuba

The process of abolition of slavery in Cuba bears some resemblance to the one in the United States, in that it was strictly related to a major conflict—the Ten Years' War (1868–1878)—which developed also as a civil war that divided the island in two halves. Previously, there had been little prospect for the end of Cuban slavery. In fact, until 1865, Cuba, still a colony of the Spanish empire, had been the constant object of U.S. southern planters' plans for annexation, while both during and after the American Civil War, Confederate planters fled to the island, which was the closest large-scale slave society still functioning at the time.[43]

Still, Cuban slavery had its days numbered, as the rise of the first Spanish abolitionist society in 1864, and then the Spanish government's abolition of the Atlantic slave trade in 1867 clearly showed. It was, however, the creation of the first Spanish republic in 1868 that precipitated slavery's fate on Cuba. Shortly afterward, a creole planter named Carlos Manuel de Cespedes launched the movement for Cuban independence. Although he freed his slaves and asked them to join the cause of Cuban independence, Cespedes was hardly an abolitionist. Yet, in order to achieve the goal of independence, by 1869, he and the planters of the eastern part of the island—soon renamed *Cuba libre* (free

Cuba)—were prepared to promise reforms leading to the abolition of slavery, provided that masters were given compensation in some form. In contrast, the planters in the western part of the island remained loyal to Spain.[44]

Finally, in 1870, in order to curb the separatist effort, the Spanish government unilaterally declared the abolition of slavery with the Moret Law. According to the law, both the children born of slave mothers since September 1868 and slaves 60 years old or more were free. Similarly to the case of the United States, this outcome was as much a victory of abolitionist politics as a result of the fact that, by fleeing the plantations and also joining the rebel army, in the words of Matt Childs and Manuel Barcia, "Cuban slaves helped transform a separatist struggle led by white slaveholding elites in eastern Cuba into a war for personal liberation."[45]

As a result of the Moret Law, the number of slaves in Cuba dropped quickly and dramatically—from 363,288 in 1868, they were reduced to 100,000 by 1883. The complete and final abolition of Cuban slavery, however, was not sanctioned until 1886 and depended in practice on the 1880 passage of the Patronato Law. This law provided for the legalization of the transition from slave labor to free labor in the form of an extended apprenticeship system, similar to the one in use in the postslavery British Caribbean, whose definitive end came only six years later, in 1886.[46]

The End of Slavery in Brazil

There is little in the process of abolition of slavery in Brazil that resembles the process of emancipation either in the United States or Cuba. In fact, in comparison, Brazilian slavery ended in a relatively peaceful way. Even though between 1864 and 1870, the country was engaged in a major conflict—the War with Paraguay—this did not have the same direct impact on slavery that the Ten Years' War did in Cuba, let alone the American Civil War in the United States.

Unlike Cuba, Brazil was an independent empire since the 1820s. However, similarly to Cuba, slavery was so much embedded in Brazilian society that an abolitionist movement did not exist until much later than in the United States. British pressures on the Brazilian government were a major factor in this respect, and they had led already to the abolition of Brazil's Atlantic slave trade in 1850.[47]

Equally important was the slaves' continuous unrest, which, in the words of Robert Slenes, "sharpened masters' long-standing fears of 'Haitianization,' already made acute by the 1835 Malé rebellion." Also, both the emperor Dom Pedro II and some sections of Brazil's elite were particularly receptive to liberal ideas of reform, including the concept of gradual abolition. As a consequence, starting from the mid-1860s, the imperial government began enacting a series of acts against slavery, culminating in the 1871 Rio Branco Law. Similar to the Moret Law in Cuba, the Rio Branco Law in Brazil freed the children born of slave mothers after 1871. However, their owners could either choose to use their services until their twenty-first year of age or accept government compensation.[48]

Predictably, the provincial slaveholding elites of Brazil, especially in the booming coffee-producing regions centered around Rio de Janeiro and São Paulo, opposed gradual abolition and the Rio Branco Law. However, the measure of actual change continued to be minimal for a number of years; in practice, the number of slaves freed by 1885 was only around 22,000 out of 1.5 million.[49]

It was only in the 1880s, with Joaquim Nabuco's 1880 foundation of the *Sociedade Brasileira Contra a Escravidão* (Brazilian Antislavery Society), that the movement for the abolition of slavery became widespread. By then, on one hand, international pressure was reaching an all-time high, since Brazil was the only large-scale slave society still in existence in the Americas. On the other hand, planters in the northeastern sugar regions were beginning to look for alternatives to slavery in the shape of sharecropping and tenancy.[50]

First, starting with Cearà in 1884, different provinces in Brazil took the initiative of abolishing slavery. Then abolitionist organizations sprang up in São Paulo and Rio de Janeiro. Meanwhile, the imperial government passed the Dantas-Saraiva-Cotegipe Law, which freed slaves who were 65 years of age in 1885. At the same time, from 1886, an increasingly massive number of runaway slaves started to leave plantations and farms, taking matters in their own hands and contributing in a major way in forcing the issue of emancipation.[51]

Finally, the imperial government acknowledged the de facto unmanageability of the situation and proclaimed the immediate end of slavery throughout Brazil with the Golden Law of 1888. It is interesting to notice that, in comparative perspective, Brazil's 1888 Golden Law, similar to the U.S. 1863 Emancipation Proclamation, was a governmental decree that freed the slaves immediately and without any form of compensation for slaveholders. Though clearly the result of two very different processes, the two comparable cases are unique in nineteenth-century slave societies in the Americas.[52]

Emancipation from Serfdom in Romania and Russia

During the same period in which the process of abolition of slavery took place in the last slave systems in the Americas, a parallel process leading to the abolition of serfdom occurred in eastern Europe. By 1860, similarly to the way slavery in the New World was restricted mainly to the three large-scale slave societies of the antebellum U.S. South, Cuba, and Brazil, serfdom was almost exclusively present in Romania and Russia.[53]

In a somewhat speedier turn of events than the one that characterized slavery's abolition in the Americas, within the space of five years, serfdom was entirely eradicated—legally, at least, by official governmental decrees—from both eastern European countries. Also, in both cases, the process of abolition of serfdom in itself occurred relatively peacefully and more as the abolition of slavery occurred in Brazil, rather than in the United States or Cuba.

On one hand, though, it is imperative to take into account the impact of widespread and ever-increasing peasant unrest, comparable to slave unrest in the Americas. On the other hand, even though it had little direct effect on serfdom, the 1854 through 1856 Crimean War had important indirect repercussions in eastern Europe in terms of weakening autocratic governments. First and foremost, it weakened Russia, whose imperial ambitions were halted, and its institutions, including serf-owning.[54]

Also as a result of Russia's defeat in the Crimean War, Russia's protectorate over Wallachia and Moldavia ended, and in 1858, the two regions joined to form the state of Romania. This is the background against which the events leading to the coup orchestrated by enlightened Prince Alexander John Cuza in 1864 took place. Once in power, Cuza dismissed the Romanian national assembly, formed by reactionary noblemen. Then, after holding a plebiscite, he issued a law that, in the words of Michael L. Bush, granted "the Romanian peasantry not only free status [thus, ending officially serfdom in Romania], but also freehold rights to the lands they farmed."[55]

The fact that the Crimean War ended with the defeat of Russia was a major factor in mobilizing Russian public opinion against serfdom. Many enlightened members of the Russian elite sympathizing with western ideas thought of serfdom as the main obstacle to a number of issues related to modernity, including modern warfare. In truth, previous czars, including Nicholas I (1825–1855), and Russian governments had debated for a long time about what they termed the "peasant question."[56]

As early as 1848, the Russian state had granted both state-owned serfs and landlord-owned serfs the right to purchase land for private ownership. However, it was only with the accession of the subsequent Czar Alexander II (1855–1881) that serious efforts to create a legislation for freeing serfs produced a material result beyond the proliferation of debates and committees. Interestingly, an indication of the seriousness of the effort may come from the fact that, in the 1850s, the Ministry of State Property conducted experiments in several estates, in regions characterized by both intensive and nonintensive agriculture, which ultimately ascertained that hired labor was comparatively more efficient and better suited than *corvée* labor for the needs of modern agriculture. According to Vera Tolz, the emancipation of serfs was part of a wide-ranging program of modernizing reforms and government policies that, under Alexander II, aimed at "enhancing nation-building" through "both ethnic and civic homogenization of the empire."[57]

The czar promulgated the law that abolished serfdom in Russia on February 19, 1861, freeing at once more than 20 million bondsmen. Extremely long and immensely complicated, the legislation created new officials called *peace mediators,* who even though serf-owners themselves, were supposed to help peasants in their transition from slavery to freedom. This transition was to be gradual, rather than sudden.[58]

As Peter Kolchin remarked, "peasants received their 'personal freedom' at once ... but they remained under the 'estate police and guardianship' of their former owners," to whom they continued to owe services in the form of *barshchina* or *obrok.* Within two years from the 1861 decree, the *pomeshchiki* (landlords) were to draw charters detailing the nature of the ex-serfs' obligations and of their land allotments. At the end of the two years, house servants were free with no land. All other ex-serfs could become free from their temporary obligations only by paying for their land allotments, and therefore becoming proprietors, in a process called *redemption.* Thus, similar to most other situations of transition from unfree labor to freedom, but in complete contrast to the 1863 slave emancipation in the United States and the 1888 slave emancipation in Brazil, in Russia, ex-unfree laborers were still bound by the law to provide additional work for a number of years as a form of compensation to their ex-owners.[59]

There is little doubt that, as Michael L. Bush has written, "the coincidental termination of New World slavery and European serfdom" was "a major event in the history of modern servitude." From our perspective, though, comparison between the two concurrent processes of abolition of unfree labor shows that slave emancipation in the United States shares only a few characteristics with both, while it has, for the most part, unique features.[60]

The main similar characteristic is the fact that, even though recent scholarship has placed correctly much more emphasis on the slaves' agency in the process of emancipation in the United States, as also in the case of Brazil and Cuba, the actual document that decreed the end of slavery in the U.S. South was a proclamation released by the official national government. This was no different from all the cases of abolition of slavery and also of abolition of serfdom. Yet, the U.S. Emancipation Proclamation of 1863 appears completely different in content from most other decrees emancipating slaves or serfs for two particular reasons:

- It freed the slaves through a war measure, both immediately and permanently.
- It provided no compensation for slaveholders.

Conversely, with the exception of Brazil's 1888 Golden Law, the other nineteenth-century emancipation decrees, including those emancipating serfs, always gave guidelines for a process of gradual emancipation, whose main purpose was to provide for some form of compensation for the former owners of unfree laborers, as Stanley Engerman has recently shown. Thus, in comparative perspective and in a wider Euro-American framework, the 1863 U.S. Emancipation Proclamation and the subsequent 1865 Thirteenth Amendment to the Constitution appear exceptionally significant because of their particularly radical provisions.[61]

In sum, the comparative perspective reinforces the idea that, in releasing the U.S. Emancipation Proclamation, in many ways, Lincoln committed an unprecedented revolutionary act. This is especially apparent when we think about it in the context of the antebellum and early Civil War policy of the Republican Party that he headed, which was both moderate and mildly antislavery.

MAKING NATIONS IN THE EURO-AMERICAN WORLD

When we turn to the actual process of nation-building that occurred through the American Civil War, we cannot help but notice that, until relatively recently, the best-known studies by historians and sociologists on the phenomenon of construction of nations in the central decades of the nineteenth century tended to focus on Europe, and sometime Asia or Latin America, mostly leaving aside the United States. Only in the past two decades, also as a result of the transnational turn in American historiography, American historians such as Thomas Bender and Carl Guarneri have turned their attention to the international context of the nineteenth-century formation of the American nation and connected it with the historiography on contemporary European nationalist movements. At present, there is a small but steadily increasing number of studies that look at the United States in its transnational context, focusing specifically on the idea of the nation and its use/reception on both sides of the Atlantic.

However, there is still too little in terms of actual comparative studies of ideas and practices of nation-building in the United States and Europe in the first half of the nineteenth century. Valid hints at possible comparisons in this sense come from the studies of David Potter and Carl Degler, who, in the 1960s and 1970s pioneered comparative investigation of the United States in the antebellum and Civil War era. Carl Degler's analysis—and later, the studies of scholars who came in his wake—dealt mainly with the Civil War as a military event that could be fruitfully compared with the wars for German unification.

However, in regard to ideology, Degler had to admit that "historians of the United States have not liked to compare Bismarck and Lincoln," as a result of the former's illiberal tendencies. On the other hand, David Potter's research focused on the political culture expressed in Lincoln's and the Republican Party's view of the nation as akin to the liberal strand of nationalism that pervaded contemporary Europe. In this sense, Potter found that a better match for a comparative study would have been Italy in the age of *Risorgimento* (which indicates the movement for Italian national unification), since

Italy was the nation whose 1861 creation signified more than any other the victory of liberal nationalist principles.[62]

National Consolidation and the Invention of Traditions

If we were to follow David Potter, nationalism and U.S. and Italian national consolidation from the point of view of political ideology and culture would be the central feature of our comparative study. Then, it would be important to begin with the most useful way to define what national consolidation implied in this sense, with specific reference to Civil War America and *Risorgimento* Italy.

From the 1980s onward, scholars of nationalism have mostly agreed with the idea—first argued by Benedict Anderson, Ernest Gellner, and Eric Hobsbawm—that modern nations are a product of different factors related to the rise of modernity. Particularly influential in this respect has been Hobsbawm's claim that modern nations bear the mark of "invented traditions." By this, he meant that nineteenth-century elites had been responsible for creating national identities in different countries through "exercises in social engineering ... often deliberate and always innovative, and aimed at planting the seeds of a national consciousness among the majority of the people." Even though Hobsbawm did not refer specifically to the American case, the implications of his assumptions were not lost on American scholars. In time, these scholars applied successfully some of Hobsbawm's views to the study of subjects as diverse as the antebellum United States and both Union and Confederate nationalisms, with works especially by Susan-Mary Grant, James McPherson, and Drew Faust.[63]

Nowadays, it is easy to see how the post-1980 scholarship on nationalism has influenced the type of comparative treatment we might find in the best transnational syntheses of U.S. history. However, scholars have yet to take into account the full implications that Hobsbawm's idea of invention of tradition might have on a comparative study focusing on the early American Republic, especially one that would pitch it against the synchronic case study of *Risorgimento* Italy.

Slavery was the main issue that deeply divided the early American republican nation economically, socially, and politically from its inception. Before and during the American Civil War, Republican nationalists supported the idea of a nation that contemplated the political possibility of an antislavery government heading the United States as a whole. This was nothing less than a novel invention of tradition in nineteenth-century America. In comparable terms, the ideas supported by Italian Moderate Liberals during the process of Italian national unification—the ideas of political independence and extension of liberal institutions to the entire peninsula—were an equally novel invention of tradition in a country such as nineteenth-century Italy, whose different areas had long been dominated by foreign or autocratic rule, or by both.

What did these two novel invented traditions truly have in common? In practice, this comparative perspective confirms that the issues that Eric Foner mentioned as specific to American nation-building in the central decades of the nineteenth century—political democracy and human liberty—were, in fact, unique in the context of the Euro-American world. Even the advanced experiment, by nineteenth-century European standards, of nation-building in Italy had as an actual outcome nothing more than a constitutional monarchy with the guarantee of basic civil rights for a minority of the population.

At the heart of the two parallel and contemporaneous novel processes of nation-building in the United States and Italy were ideologies that had profoundly different concepts of liberty: one more universal and the other more restricted. Yet, within their respective context, these two ideologies contained novel elements that advanced, in different ways and degrees, the general cause of freedom and progress, and therefore constituted a rupture with a more conservative past. This is the only possible way to make sense of David Potter's 1968 arguments. According to Potter, "the uprising of the North," under the guidance of Lincoln's Republicans, "in 1861, coming in the same year in which Victor Emmanuel was crowned king of a united Italy, marked the turning of the tide which had been running against nationalism for the preceding forty-five years," and "it forged a bond between nationalism and liberalism at a time when it appeared that the two might draw apart and move in opposite directions" after the defeat of the 1848 European revolutions.[64]

Interestingly, this is the transnational theme explored by Tim Roberts in his book *Distant Revolutons* (2009), in which he shows how aware American antislavery politicians were of the link between Europe's movements for national freedom and their own struggle to free the American nation from slavery. Not accidentally, in thinking comparatively, Potter had come to the conclusion that the United States and Italy shared a phenomenon of national consolidation in which a comparable type of liberal nationalism, based on different but related principles of freedom and progress, had prevailed. [65]

Following in Potter's footsteps and keeping his framework as a valid basis of comparative analysis, we can trace the development of the American and Italian invented traditions of nationhood through a synchronic comparative study of the trajectories followed by the process of formation and reconstitution of the two nations. The point of departure is the investigation of how two different processes of nation-building, occurring in two entirely different settings, were partly guided by comparable ideological principles: Republican nationalism in the United States and moderate liberalism in Italy. Ultimately, both processes saw the establishment, or reestablishment, through war, of a unified nation along those principles. In an attempt to understand comparatively the reason for the occurrence of national consolidation in the two countries along those lines—two parallel processes in which Lincoln and Cavour played crucial roles—we need to first focus on the crucial decade of the 1860s.

Republican Nationalism in the Civil War United States

Since the time of the "House Divided" speech, the task that Lincoln had set for himself and the Republican Party was to maintain the integrity of the American nation, even at the cost of fighting a civil war to block the slave South's attempt to secede with the formation of the Confederacy. This happened shortly after he was elected President of the United States in November 1860. In practice, during the Civil War, Lincoln acted on his project, which was one of national reunification of the American nation along liberal principles. Through the war, he tested the strength of this idea, and in the process changed and radicalized it.

In 1861, Lincoln's ideology was one of moderate Union-focused nationalism, through which he opposed slavery's expansion. He did not interfere with slavery and the domestic institutions of southern states, because it would have been unconstitutional. With the war, that ideology changed when Lincoln came to place the very idea of unconditional liberty, for whites and blacks alike, at the heart of the struggle.[66]

Lincoln's Republican nationalism was based on the actual application of the principles of the Declaration of Independence, rather than on an abstract allegiance to them. Lincoln's vision for the new American nation that would emerge from the war relied on both his own deep personal beliefs, as they had continuously evolved since he had started his political career, and on the ideas he had matured in the course of the Civil War.[67]

In the midst of the process of forging this "new national patriotism," as Melinda Lawson has noted, Lincoln asked his countrymen's allegiance to a nation—the Union—that was to be "strong and beneficent, bestowing economic well-being and guaranteeing liberty to its people." Though at the beginning of the war, in 1861, he had no intention to touch slavery, in his program for national reunification, Lincoln clearly envisioned a type of American nation that opposed in principle everything the Confederate States of America stood for, namely limitation of the federal government and slavery. Three years on, in 1864, in reviewing the causes of the Civil War and the path taken by his commitment to nation-building along these lines, Lincoln wrote that "we accepted this war for ... a worthy object, and the war will end when that object is attained ... [it] has taken three years, it was begun or accepted upon the line of restoring national authority over the whole national domain."[68]

A year earlier, in the 1863 Gettysburg Address, Lincoln's vision for the American nation had found its most complete expression. In it, Lincoln reminded Americans of how the Revolution had founded "a new nation, conceived in Liberty, and dedicated to the proposition that all men are created equal" and proclaimed solemnly how "this nation, under God, shall have a new birth of freedom."[69]

In envisioning the new American nation that was to emerge after the Civil War, Lincoln reaffirmed the validity of the liberal principles of the Declaration of Independence as its foundation stone and reiterated his own interpretation of the American Civil War as a struggle for national reunification along not just Republican nationalist, but truly, inclusively, democratic lines. In this respect, it is possible to say that, with Lincoln's crucial contribution, the process of nation-building and reunification in the United States led to a novel invented tradition of nationhood. This tradition, for the first time, was inclusive of both whites and blacks, and that made possible, uniquely in the Euro-American world, the combination of the moderate principles of Republican nationalism with the revolutionary character of slave emancipation.[70]

U.S. Republican Nationalism and Italian Liberal Nationalism in Comparative Perspective

Despite the very different contexts and concepts of freedom and nation, we can make comparable comments in terms of novelty and invented tradition of nationhood when we turn to Italian national unification, and therefore to the process of nation-building mostly guided by Cavour and the Moderate Liberals during the Italian *Risorgimento.*

Similar to slave emancipation in the United States, national unification in Italy was a combination of factors, some of which were accidental. Even though, in 1859, Cavour's plan was for a constitutional kingdom of northern Italy, he had struck an alliance with some of the most important democratic leaders, who gathered in the National Society, which was an organization largely controlled by him and dedicated to the achievement of national Italian unity under the Moderate Liberals.[71]

Giuseppe Garibaldi was also among the National Society's members. He decided to single-handedly proceed to the military conquest of the Bourbon Kingdom in the Italian south. He achieved this by October 1860, and then handed the kingdom over to the Piedmontese King Victor Emmanuel II, thus ending Cavour's projected scenarios of an Italy divided in northern and southern sections. Shortly afterwards, Cavour spoke in Parliament in October 1860 on the subject of ratification of southern Italy's annexation. In his speech, Cavour significantly clarified that Italy's national unification had a decisively liberal and antirevolutionary character by remarking that "in the last two years, Italy has given a wonderful example of civil wisdom by her attachment to the principles of order, morality and civilization" and advocating as a witness "the impartial voice of enlightened, liberal Europe."[72]

Comparable to the process of nation-building in the United States guided by Lincoln and the Republicans, the process of national unification in Italy, headed mostly by Cavour and the Moderate Liberals, produced a novel concept of nationhood. It is possible to call this concept an invented tradition, in that it allowed the formation of a novel national political institution, in which the basic freedoms and civic rights were, for the first time in Italian history, defended by an actual constitution (the 1848 Piedmontese Albertine Statute, extended to the rest of Italy in 1861). Yet, this is where the similarities between the two case studies end.

In the United States, Lincoln and the Republican Party radicalized their idea of liberty in the course of the Civil War to the point of accepting slave emancipation. However, in Italy, Cavour and the Moderate Liberals succeeded in containing the process of national unification within liberal terms, not allowing more radical ideas or projects to contribute to the creation of a more democratic Italian nation. Therefore, in a way, the outcome of the creation of the two invented traditions of nationhood was opposite in terms of actual practical applications of freedom in civic and political terms. The United States was ultimately a largely more inclusive nation. Conversely, Italy was ultimately still an extremely restrictive nation, as was the norm in most of continental Europe and Latin America. This shows how truly outside the norm the American case study actually was.[73]

Still, it is interesting to notice that, during the early stages of the American Civil War and immediately after Italian unification, the network binding the two countries and their parallel nation-building paths was most famously tested with Abraham Lincoln's 1861 offer of the post of Commander in the Union Army to Giuseppe Garibaldi, the icon of Italian nationalism. Effectively, in unifying Italy in October 1860 in the name of the constitutional Piedmontese monarchy, Garibaldi—a republican and democratic nationalist who wound up joining the Moderate Liberals—acted in the name of comparable principles to those that were inspiring Lincoln: to build an effectively united nation in which basic freedoms were guaranteed. Besides the actual military aspects Lincoln might have considered in making his offer, it was the fact that Garibaldi had effectively created an Italian liberal nation that must have prompted Lincoln to think that the Italian general might have been able to help him create the type of American republican nation he had in mind. As Carl Guarneri has written in regard to the two protagonists of the two processes of national unification, "Lincoln saw enough similarities that he asked the Italian nationalist hero Giuseppe Garibaldi to lead a contingent of Union troops."[74]

A comparative and transnational study between the American Civil War and Italy's national unification, more than one between the American Civil War and Germany's national unification, would show its primary value in being a test of a sort. Such a test would lead to our better understanding of ideas and practices referring to what Peter Parish called "meliorative nationalism," because of the link between the ideas of progress and freedom at its core. Parish makes specific references to Lincoln and the American Republican Party, but we could equally apply the expression "meliorative nationalism" to the ideology that characterized Cavour and Italy's Moderate Liberals—effectively, the political movement that guided Italy's process of unification.[75]

In short, through this specific comparison, it is possible to investigate the practice of what several nineteenth-century observers thought of as two parallel exercises in progressive nation-building on the two sides of the Atlantic in the mid-nineteenth century. In turn, such a study may turn out particularly beneficial in showing how mid-nineteenth century American politics were also part of a much wider context—one that was as much Atlantic as genuinely Euro-American. In this sense, the United States and Italy were but two of many different parts of a nineteenth-century political milieu that was much more global than we once thought, as shown by both Eric Hobsbawm in *The Age of Capital* (1975), and C. A. Bayly in *The Birth of the Modern World* (2004).[76]

At the same time, though, comparison shows clearly that the American Civil War is a unique case study of nation-building in the entire Euro-American world—a case study only partly comparable particularly with Italy's national unification. The American Civil War's unique characteristics lay in the fact of being a process of construction of a nation that occurred with the help of the very particular type of nationalist ideology of Republican nationalism, even though it included some common features with the European type of liberal nationalism that characterized Italy's Moderate Liberals.

Unlike the nationalist ideology that characterized Cavour and his allies, Republican nationalism had within itself the potential for becoming more radical. And it did, in fact, become increasingly radical during the conflict resulting from nation-building in the course of the American Civil War, to the point of including, by 1863, a major revolutionary element in the shape of slave emancipation. In this sense, comparison reinforces, rather than debunks, the very idea of "American exceptionalism" in relation to the unique coexistence in the American Civil War of an originally moderate concept of nation-building with a later radical, and even revolutionary, emancipationist idea.

The American Civil War in Its Euro-American Dimension

We can look at nationalist ideology, the making of slave emancipation, and the actual process of nation-building as three different and equally important elements of a Hegelian dialectic in understanding the American Civil War. When seen in comparative perspective in a Euro-American dimension, each of these three elements shows a few remarkable similarities and some striking differences with nationalism, emancipation, and nation-building, as they occurred or manifested themselves in other areas of the Atlantic world and beyond. Such similarities and differences help us to better place the significance of the American Civil War in its proper global context, since it is thus truly possible to grasp its unique character.

To begin with the first element of the Hegelian dialectic, the *thesis,* comparison with both Europe and Latin America shows that the Republican Party's and Abraham Lincoln's focus on the importance of the American Union as the defender of the principles of freedom and progress has something in common with the type of liberal nationalism that characterized liberal nationalist movements and liberal ideologies in most of the nineteenth Euro-American world. This accord is in terms of the emphasis on the indissoluble link between national progress and basic civil rights, and also in terms of the nonradical character of its demands.

The second element of the Hegelian dialectic—the *antithesis*—is represented by slave emancipation in the United States. Comparison with the nineteenth-century processes of abolition of slavery in the other two large-scale slave societies in the Americas and of abolition of serfdom in eastern Europe shows that U.S. slave emancipation occurred with characteristics that make it similar only to Brazilian slave emancipation, as some historians have already noticed. The 1863 Emancipation Proclamation decreed, with a war measure, immediate and permanent freedom for the slaves, without any compensation for slaveholders, These characteristics are mainly related to the much more radical departure that the Emancipation Proclamation represented in comparison with both the mild antislavery policy of the antebellum Republican Party and the processes of gradual emancipation occurring in most of the nineteenth-century Euro-American world.

We would be entitled to see these two elements—the moderate type of Union-focused Republican nationalism with a noninterventionist antislavery component and the radical step of emancipating slaves—as antithetical, as they were to a certain extent. However, we should also acknowledge the fact that the actual process of nation-building in the United States through the American Civil War—the third element in Hegelian dialectic, or *synthesis*—succeeded in creating the conditions for their harmonious coexistence in a way that is not comparable to that of any other process of construction of nations in the Euro-American world. This coexistence, though, could not have been possible without the particular characteristics of the Republican nationalist ideology of the Republican Party. Within that ideology was the potential to transform a moderate appeal to the deepest American national identity into a radical message of inclusive freedom, if the circumstances of nation-building turned out to be propitious for such transformation, as they actually did under the skillful guidance of Lincoln.

In this sense, the ideology that characterized Lincoln and the Republican Party by 1863 represented a new departure in the idea of American nationhood. This ideology, even keeping in mind the crucial differences in contents and contexts, is only very generally comparable, in terms of being a novel invention of tradition, with the new departure that Cavour and the Moderate Libcrals represented in Italy in the same years. In fact, in the United States, this new departure created the fundamental preconditions for the making of the exceptional character of the American Civil War as a process of nation-building. This was a process that joined together, uniquely in the Euro-American world, the moderate brand of Republican nationalism with the revolutionary act of slave emancipation.

CONCLUSION

AMERICAN SLAVERY IN ATLANTIC AND EURO-AMERICAN PERSPECTIVE

Current comparative studies on American slavery have moved toward a better appreciation of the New World and Atlantic contexts of the rise and fall of the "peculiar institution" in the United States. Nowadays, the best available research focuses increasingly on the continental New World perspective on American slavery in comparison with either Latin America or the Caribbean. Many of these scholarly studies pay particularly close attention to American slavery's transatlantic links with Africa through the Atlantic slave trade, in the same way that scholarship on other slave societies in the Americas looks for similar links. At the same time, more scholars are now investigating comparative themes across the slave societies that underwent the transformations related to the rise of the second slavery in the Atlantic world.

However, few studies have broadened this Atlantic perspective and attempted to engage in sustained comparison between American slavery and European serfdom or other forms of labor. And yet, servitude in many different forms was common in large areas of the world until the late nineteenth century. Therefore, on one hand, in order to understand correctly the world context in which the rise and fall of American slavery occurred, we need to compare it with other forms of free and unfree labor, such as the ones that characterized eastern and southern Europe. On the other hand, the Euro-American context, through comparison between progressive abolitionist and nationalist movements that sprang up on the opposite sides of the Atlantic, also helps us understand the significance of American traditions of antislavery, and ultimately, of the making of emancipation in the course of the American Civil War.

AMERICAN SLAVERY IN ITS ATLANTIC CONTEXT

When looking at a historiographical overview of the scholarship on comparative slavery in the Americas, beginning from Frank Tannenbaum's *Slave and Citizen* (1946) and ending with Robin Blackburn's *The American Crucible* (2011), we notice an interesting feature. In the 65 years from 1946 to 2011, slavery studies in the United States have experienced something akin to a "paradigm shift"—in the definition of science historian Thomas Kuhn, from American slavery to Atlantic slavery. In practice, they have moved away from their previous focus on the colonial and antebellum American South in comparison with specific slave societies in the New World and toward an awareness of the particular position of American slavery within the American continent, and also within the broader context of the Atlantic world, encompassing not only the Americas, but also Africa and Europe.[1]

When seen from a continental New World perspective, the fact that only 5 to 6 percent of the slaves transported from Africa to the New World went to the North American colonies, as opposed to the almost 90 percent that went to Brazil and the Caribbean, make especially the beginnings of American slavery in the seventeenth century look definitively peripheral within the Atlantic slave system. Yet, already by the early eighteenth century, American slavery was developing unique characteristics, especially in regard to the slaves' highest rate of self-reproduction anywhere in the Americas, which were to bear crucial consequences in the future. At the same time, the most recent scholarship has demonstrated that the Atlantic context—particularly the study of connections with contemporary Europe and Africa as well as those within the American continent—is particularly important in understanding the ideological elements that American slaveholders shared with other colonial elites, and also the experience of enslavement and bondage that African American slaves shared with other enslaved African populations in the Americas.

The Atlantic context has also proven helpful in gaining a broader understanding of the emergence of specific American systems of racial slavery and of their increasingly strict legal definitions. These systems followed different paths and moved at different paces in the particular New World slave societies, all unique in their own way in the Euro-Afro-American cosmos. Especially toward the end of the eighteenth century, similarly to the other slaveholding elites of the New World, American planters were heavily engaged in an elite reformist culture, which was spread throughout the Atlantic with different adaptations to the local settings. Broadly speaking, in all the different settings, the elite reformist culture focused particularly on attempts at rationalizing both agricultural production and slave management.

But by the later part of the eighteenth century, under the effect of momentous events in both North America and Europe, the Atlantic world was undergoing an Age of Revolutions, which slaveholders and slaves interpreted in radically different ways. To the reformist and relatively conservative revolutions prompted by the slaveholding elites throughout the New World in the period of 1770 to 1830, the slaves responded with a unique type of ultra-radical revolution that stemmed directly from events in France and that eventually created, unexpectedly, the Haitian Republic in the French Caribbean colony of St. Domingue.

The end of the most profitable slave society and colony in the entire Atlantic and the subsequent economic disruption forced all the slaveholding elites of the New World to envision and act upon a radical adjustment of the Atlantic slave system. At the same time, the Haitian Revolution continued to bear an enormous influence throughout the Atlantic world as a permanent example of the only successful slave revolt, particularly in the early decades of the nineteenth century.

Within the context of the nineteenth-century Atlantic world, a broad comparative perspective allows us to gain insights on the perennial question of whether the nineteenth-century U.S. South was a capitalist or precapitalist economy/society. In other words, how "modern" was nineteenth-century American slavery, when seen in comparative perspective?

In comparison with other neighboring and flourishing slave societies in the Caribbean and Latin America, such as Cuba and Brazil, which underwent similar transformations with the rise of the second slavery in the nineteenth-century, the antebellum U.S. South certainly seems both modern and capitalist. We can draw this conclusion particularly through the study of the technological and biological innovations that accompanied the focus of the plantation systems of the U.S. South, Cuba, and Brazil on those cash crops that commanded the world economy (cotton, sugar, and coffee), and from the study of the slaveholders' type of elite culture and attitude toward innovation and efficiency *vis-à-vis* both agricultural production and slave management.[2]

The general impression that modern features characterized the antebellum U.S. South, Cuba, and Brazil under the second slavery is further reinforced by comparison with the Sokoto Caliphate, the only other comparable large-scale slave society in the nineteenth-century Atlantic. This was a society in which agricultural slavery and the plantation system were the most important sectors of the economy, but the ties to the world market and the consequent need for efficient production and technological innovation were minimal.

For their part, the slave cultures and the traditions of resistance in the antebellum U.S. South, Cuba, and Brazil appear to have been unique, even though remarkably different among themselves, in many respects. On one hand, the syncretism that characterized Afro-Christian slave religion in all three slave societies occurred in different ways and degrees, but in all the cases gave origin to entirely novel cults. On the other hand, religion remained, for the slaves—again in different ways and degrees—the main way of maintaining contact with the surviving features of their distinct African identities.[3]

In the long run, the contact of different African traditions with different New World settings and white cultures led to a process of creolization, and to the birth of distinctive cultures, at once both African and American. This becomes clear particularly in studying the role of religion in several examples of slave revolts. Also in this case, comparison with the Sokoto Caliphate as another major slave society in the Atlantic, and one with a completely different culture, is illuminating, since the importance of religious identity, particularly Muslim identity, among the slaves is also evident in the episodes of slave revolt. The fact that a number of Muslim slaves from Sokoto ended up in Brazil, causing one of the largest slave revolts in the New World—the Malé rebellion in Bahia in 1835—shows the crucial importance of the Atlantic context in understanding the links between the nineteenth century's largest slave societies.

THE IMPORTANCE OF THE EURO-AMERICAN CONTEXT

Despite the popularity of the paradigm of Atlantic slavery in current studies on the U.S. peculiar institution in broader perspectives, several scholars have already begun to move beyond its relatively narrow geographical scope, which encompasses both the Americas and Africa but only Atlantic Europe, and its equally narrow focus on slavery as a unique form of unfree labor. Combining these studies with my own research, I have argued that, in order to truly understand the historical significance of American slavery, particularly in the nineteenth century, we need to tie the Atlantic context to the Euro-American world—a world that encompassed equally the whole of the Americas and the whole of Europe and the Mediterranean. In practice, I believe we should engage in sustained comparison of American slavery with various forms of European servitude and nominally free labor. Specifically, drawing from key studies that use a rigorous comparative approach, I have given some examples of comparison between the ideology and practice of antebellum American slavery and the ideology and practice of eastern and southern European forms of labor in the nineteenth century—specifically, Russian serfdom, Prussian semi-serfdom, and southern Italian and Spanish sharecropping and tenancy.

As we have seen, within the broader context of the Euro-American world, there existed a number of different shades of freedom and unfreedom in relation to the work performed by agricultural laborers. These specific comparisons also help us to further understand the true meaning of the connection between modernity and slavery, or rather between modernity and forms of free and unfree labor. In this comparative exercise, the analysis of the variety and, to a certain extent, the relativity of the strict legal definitions of *free* and *unfree* labor when referring to Euro-American systems of agricultural work play a crucial role. Equally important is the ideological correspondence between agrarian elites in the Americas and Europe in respect to themes such as agronomy, scientific agriculture, and improvement in the efficiency of management of both land and workers, free and unfree.[4]

Yet, the story of American slavery in comparative perspective is also the story of the demise of a powerful slave system. This demise would not have been possible without the existence of abolitionist movements, as David Brion Davis, Seymour Drescher, and Robin Blackburn have all recently reminded us. True to the paradigm of Atlantic slavery, current scholarship on American abolitionism sees this as part of a large-scale and ongoing common struggle, joined by the British and the French, to bring down the Atlantic slave system, debating mostly whether the struggle occurred for economic or moral reasons.[5]

There is no doubt that comparison between the different abolitionist movements that operated at different times in different countries is very useful. However, it is more important than has been so far acknowledged to look at the Euro-American context of Atlantic abolitionism in order to add a further, crucial dimension through comparative analysis.

Abolitionism had the radical idea of immediate emancipation of the slaves at its core. Democratic nationalism maintained the idea of immediate liberation of oppressed nationalities. Comparison between the United States and Britain on one hand, and Europe on the other hand shows that Anglo-American abolitionism, and partly French abolitionism, played in the Atlantic world an ideological role similar to the one that democratic nationalism played in the European continent. In particular, comparison between Atlantic abolitionism and Irish and Italian nationalism, with a specific focus on the renowned historical figures of Atlantic progressives William Lloyd Garrison, William Henry Ashurst, Daniel O'Connell, and Giuseppe Mazzini, highlights the ideological links between different types of nineteenth-century radical activists. This, in turn, suggests possible new departures for the study of the American abolitionist movement within the larger context of nineteenth-century radicalism and in comparative transatlantic and Euro-American perspective.

Ultimately, though, emancipation and the destruction of American slavery were the consequences of the Civil War. It is now commonplace to state, in somewhat implicit comparative fashion, that the United States became a real nation with the American Civil War, but what does this mean in relation to the U.S. peculiar institution—the primary cause of the war? Here, comparison with how emancipation occurred in other advanced Atlantic slave societies, such as Cuba and Brazil, as well as with how it occurred in Eastern European serf societies such as Russia, helps us understand the uniqueness of the American case. Note that this was the only case, aside from Haiti, in which immediate and permanent emancipation occurred by force of arms and without compensation for slaveholders.

However, to understand clearly the crucial link between the demise of American slavery and the Civil War as parts of a nation-building process, we must focus more closely on the Euro-American context. Recent studies by different scholars—particularly those by C. A. Bayly, Thomas Bender, and Nicholas and Peter Onuf—have pointed out the ideological similarities between mid-nineteenth-century experiments in nation-building on the two sides of the Atlantic, in the Americas and Europe. These experiments focused on the combination of nationalist movements with either the achievement or the acknowledgment of basic civil liberties. Yet, it is also particularly important to see how the actual modes of expression of the conflict between slavery/oppression on one side and freedom on the other side, and also of the resolution of that conflict, specifically underpinned the ideological justification of comparable cases of contemporary refoundations of both an American nation emancipated from slavery and of European nations emancipated from national oppression, particularly Italy, in the decade 1860 to 1870. In this respect, comparison between the processes of nation-building and slave emancipation through the American Civil War and the making of the Italian monarchy, which was nineteenth-century Europe's most successful case of liberal nationalism, in Euro-American perspective adds a further dimension to the study of the U.S. peculiar institution and its rise and fall within a much wider context.[6]

POST-EMANCIPATION SOCIETIES

Within the wider context of the global process of emancipation from slavery, comparison shows that, as Robin Blackburn has remarked, "bourgeois abolitionism became an effective emancipatory force only where it was prepared to reach a 'historic compromise' with broader social forces, including slaves, free people of color, white artisans, immigrants and the native-born ... [by] countenancing inroads on already established property rights and extending the bounds of citizenship." This happened in 1794 and 1848 France, in 1833 Britain, in Civil War America, and in 1888 Brazil. Yet, uniquely within the Euro-American world, the revolutionary dimension of U.S. slave emancipation led to subsequent particularly revolutionary transformations in the concept of national citizenship during the period of Reconstruction (1865–1877).[7]

Specifically during the phase of Radical, or Congressional, Reconstruction, the Radical Republicans who dominated Congress and national politics redefined the concept of national citizenship by acting further on the idea of inclusive freedom, not only through the official sanction of the end of national slavery in the United States with the passing of the Thirteenth Amendment (1865), but also through specific legislation that protected the civil rights, and particularly the right to enfranchisement, of the newly freed African American population with the passing of the Fourteenth and Fifteenth Amendments (1868 and 1870). Yet, as Eric Foner has remarked, "when it came to the former slaves' quest for land ... Reconstruction governments took few concrete actions." As a result, unable to own land, most African American families rented it from the planters through a system of sharecropping, in some ways comparable to the one in use in areas of southern Italy and Spain. This system perpetuated the ex-slave's dependent status, and, eventually, was accompanied by a gradual loss of civil rights and increasing racial discrimination in the later quarter of the nineteenth century.[8]

The level of racial discrimination and segregation that African Americans experienced under Jim Crow, starting from the last decades of the nineteenth century, provides the United States with another unique characteristic in historical terms. As Robin Blackburn has noted, unlike what happened in the United States, "in the decades immediately succeeding emancipation the freedmen and women of Jamaica, Brazil, Cuba, and Haiti experienced substantial material and spiritual benefits from their new conditions." Certainly, a major reason was the absence of a strict racial regime in these four regions in comparison with the United States.[9]

On the other hand, a common negative feature throughout the Americas was the chronically low amount of land owned by ex-slaves and their descendants. Such a situation might prove conducive to major uprisings, as in the case of the 1865 Morant Bay Rebellion in Jamaica. Ultimately, in all the post-emancipation societies, it led to the perpetuation of a social order at whose bottom continued to be the free persons of color, even after the proclamation of the republic in Brazil in 1889 and the end of Spanish colonial rule in Cuba in 1898. Thus, there is no doubt that, as Edward Rugemer has remarked, "as [Charles] Sumner [one of the leaders of Radical Reconstruction in the United States] and the rebels at Morant Bay understood, the destruction of slavery led to new forms of exploitation that impoverished the freedom of emancipation."[10]

Yet, in the longer term and in transatlantic perspective, the actual destruction of slavery took an even lengthier time in the case of the Sokoto Caliphate. Its definitive collapse occurred only in 1903, as a result of British, French, and German appropriation of different parts of its territory. Even though keen to destroy the Caliphate's power, the European colonial empires deliberately maintained its economic system based on slavery, since only enslavement and the slave trade were prohibited, while slavery itself continued in the areas under British rule until as late as 1936.[11]

Comparably to the long end of slavery in the Atlantic world, the long end of serfdom in eastern Europe did not lead to a sudden end in the power of the agrarian elites. In comparative perspective, Europe's emancipated serfs, similar to emancipated slaves in the Americas, struggled to become landed proprietors. For example, in Russia, the process of "redemption" initiated by the 1861 laws on serf emancipation allowed former serfs to become free from obligations once they were able to purchase the land they worked on. However, according to Peter Kolchin, it took 20 years for the majority of the peasants to become proprietors, since only by 1881 did four-fifths of them finally own land.[12]

Throughout the nineteenth century, the chronic lack of landownership also affected in equal ways the long-emancipated peasants of southern Europe, leading to an even more widespread use of sharecropping. As Carl Levy remarked, through sharecropping, "the old nobility and the upstart notables of Spain, France, and Italy could meet the needs of modern markets, while at the same time they exercised a near feudal power of oversight over extended families of sharecroppers." This situation reminds us in many ways of planters and freed people in the U.S. South after the Civil War.[13]

⟴

Ultimately, scholars who study the rise and fall of American slavery in comparative and international perspective must acknowledge even more deeply than they do now the importance of its Atlantic context, not just at the time of initial expansion of New World slave societies and at the peak of the Atlantic slave trade, when the mechanisms of empire created crucial links between the American South and other slave systems in the Americas and Africa, but also in the subsequent historical period, after the closing of the slave trade.

In the nineteenth century, the antebellum U.S. South was one of the three "advanced" slave societies interested by the phenomenon of the second slavery, together with Cuba and Brazil, with which it had a great deal in common. However, it was also one of the four largest-scale slave societies present in the Atlantic world, if we include the Sokoto Caliphate in Africa. Comparison between the antebellum U.S. South, Brazil, and Cuba is now being practiced fairly often, and it is likely to yield insights in a variety of respects. Comparison between the antebellum U.S. South and the Sokoto Caliphate is not yet an issue in contemporary scholarship. However, there is little doubt that the few interesting similarities and the many differences between the two large-scale slave societies and plantation economies would yield other types of insights altogether.

Comparative history of American slavery needs to go beyond the Atlantic context in order to truly lead to a much deeper understanding of the nature and meaning of the U.S. "peculiar institution." Only by placing American slavery clearly in an international perspective within a Euro-American context, and by comparing it with both eastern European serfdom and with different forms of nominally free labor in Mediterranean agriculture, such as sharecropping and tenancy, can we reach a true comprehension of slavery's actual position in the nineteenth-century world and in the world economy as a whole.

At the same time, we can gain further insights by placing the American tradition of antislavery and abolitionism in its Euro-American context, by linking it and comparing it with both Atlantic abolitionism and European nationalism. In this respect, the fact that there were several contacts between

American antislavery activists and European nationalists should prompt us to reflect on the parallels between the two struggles.

Finally, following in the footsteps of comparative and transnational historians who have looked at the nineteenth-century age of emancipation and nation-building in their wider world milieus, we should also consider the American Civil War and slave emancipation in the U.S. South within an international Euro-American perspective, seeking to gain insights from comparisons with contemporary movements, both for emancipation from forms of unfree labor and for the creation of progressive nations. In this respect, a Euro-American perspective on post-emancipation societies—first and foremost the U.S. South during and after Reconstruction—allows us to better understand, despite the many differences, how all the processes of transition to free labor, and their aftermath, allowed the agrarian elites of different countries to retain their power to a large extent, while also keeping the rural laborers in a condition of economic dependency and social subordination. This was mainly a consequence of the widespread lack of peasant landownership—a phenomenon that, in the later part of the nineteenth century, characterized equally the Atlantic world and Europe.

NOTES

PREFACE

1. The expression "peculiar institution"—already used by John C. Calhoun and Alexander Stephens in the nineteenth century—has been reproposed in modern scholarly studies, notably in the title of Kenneth Stampp's classic work *The Peculiar Institution: Slavery in the Antebellum American South* (New York, 1955), and has always invited comparison, even only for the purpose of establishing what exactly was "peculiar" about American slavery.

2. See C. A. Bayly, *The Birth of the Modern World, 1780–1914: Connections and Comparisons* (Oxford, 2004).

INTRODUCTION

1. Marc Bloch, "Pour une histoire comparée des sociétés européennes," *Revue de synthèse historique* 46 (1928), 15-50; and Theda Skocpol and Margaret Somers, "The Use of Comparative History in Macro-Social Enquiry," *Comparative Studies in Society and History,* 22 (1980), 174-197.

2. Peter Kolchin, *A Sphinx on the American Land: The Nineteenth-Century South in Comparative Perspective* (Baton Rouge, LA, 2003).

3. See Peter Kolchin, *Sphinx on the American Land,* 4.

4. Frank Tannenbaum, *Slave and Citizen* (New York, 1946).

5. Stanley Elkins, *Slavery: A Problem in American Institutional and Intellectual Life* (Chicago, 1959).

6. See Herbert Klein, *Slavery in the Americas: A Comparative Study of Virginia and Cuba* (Chicago, 1967); and Carl Degler, *Neither Black Nor White: Slavery and Race Relations in the U.S. and Brazil* (New York, 1971).

7. See Eugene Genovese, *The World the Slaveholders Made: Two Essays in Interpretation* (New York, 1968); and Genovese, *Roll, Jordan, Roll: The World the Slaves Made* (New York, 1974).

8. See Gwendolyn Midlo Hall, *Social Control in Slave Plantation Societies: A Comparison of Saint Domingue and Cuba* (Baltimore, MD, 1971).

9. See Gilberto Freyre, *The Masters and the Slaves* (New York, 1956, orig. pub. in 1933); C. L. R. James, *The Black Jacobins* (New York, 1938); Eric Williams, *Capitalism and Slavery*

(London, 1944); Genovese, *From Rebellion to Revolution: Afro-American Slave Revolts in the Making of the Modern World* (Baton Rouge, LA, 1979); and Laura Foner and Genovese, eds., *Slavery in the New World: A Reader in Comparative History* (New York, 1969).

10. See David Brion Davis, *The Problem of Slavery in Western Culture* (New York, 1966); Davis, *The Problem of Slavery in the Age of Revolutions, 1770–1825* (Ithaca, NY, 1975); and Davis, *Slavery and Human Progress* (New York, 1984).

11. See Orlando Patterson, *Slavery and Social Death: A Comparative Study* (Cambridge, MA, 1982).

12. See Immanuel Wallerstein, *The Modern World-System,* 4 vols. (New York, 1974–2011).

13. See Eric Wolf, *Europe and the People Without History* (Berkeley, CA, 1982).

14. See Philip Curtin, *The Atlantic Slave Trade: A Census* (Madison, WI, 1969); and Herbert Klein, *The Middle Passage: Comparative Studies in the Atlantic Slave Trade* (Princeton, NJ, 1978).

15. See Philip Curtin, *The Rise and Fall of the Plantation Complex: Essays in Atlantic History* (New York, 1990); Robin Blackburn, *The Making of New World Slavery, 1492–1800: From the Baroque to the Modern* (London, 1997); Seymour Drescher, *From Slavery to Freedom: Comparative Studies in the Rise and Fall of Atlantic Slavery* (New York, 1999); David Brion Davis, *Inhuman Bondage: The Rise and Fall of Slavery in the New World* (New York, 2006); Stanley Engerman, *Slavery, Emancipation, and Freedom: Comparative Perspectives* (Baton Rouge, LA, 2007); Drescher, *Abolition: A History of Slavery and Antislavery* (New York, 2009); and Blackburn, *The American Crucible: Slavery, Emancipation, and Human Rights* (London, 2011).

16. See Philip Curtin, *Rise and Fall of the Plantation Complex.*

17. See Robin Blackburn, *Making of New World Slavery.*

18. David Brion Davis, *Inhuman Bondage,* 2.

19. See Seymour Drescher, *Abolition*; and Robin Blackburn, *American Crucible.* See also Robert L. Paquette and Mark M. Smith eds., *The Oxford Handbook of Slavery in the Americas* (New York, 2010).

20. See Ira Berlin and Philip Morgan, eds., *Cultivation and Culture: Labor and the Shaping of Slave Life in the Americas* (Charlottesville, VA, 1993); Rafael de Bivar Marquese, *Feitores do corpo, missionarios da mente: Senhores,letrados e o controle dos escravos nas Americas, 1660–1860* (São Paulo, 2004); and Laird Bergad, *The Comparative Histories of Slavery in Cuba, Brazil, and the United States* (New York, 2007). On the concept of "second slavery," see particularly Dale Tomich, *Through the Prism of Slavery: Labor, Capital and World Economy* (Lanham, MD, 2004).

21. See James Walvin, *Questioning Slavery* (London, 1996); and Herbert Klein and Ben Vinson, *African Slavery in Latin America and the Caribbean* (New York, 2007). See also, particularly for the comparison of slave rebellions, Seymour Drescher and Pieter C. Emmer, eds., *Who Abolished Slavery? Slave Revolts and Abolitionism: A Debate with Joao Pedro Marques* New York, 2010).

22. See Barbara Solow, ed., *Slavery and the Rise of the Atlantic System* (New York, 1993); Joseph Inikori and Stanley Engerman, eds., *The Atlantic Slave Trade: Effects on Economies, Societies, and Peoples in Africa, the Americas, and Europe* (Durham, NC, 1992); Stuart Schwartz, ed., *Tropical Babylons: Sugar and the Making of the Atlantic World* (Chapel Hill, NC, 2003); David Barry Gaspar and Darlen Clark Hine, eds., *More than Chattel: Black Women and Slavery in the Americas* (Bloomington, IN, 1996); and Gwyn Campbell, Suzanne Miers, and Joseph Miller, eds., *Women and Slavery,* Vol. 2: *The Modern Atlantic* (Athens, GA, 2007).

23. ; See Paul Gilroy, *The Black Atlantic: Modernity and Double Consciousness* (London 1993); John Thornton, *Africa and Africans in the Making of the Atlantic World, 1440–1800* (New York, 1998); Stephanie Smallwood, *Saltwater Slavery: A Middle Passage from Africa to American Diaspora* (Cambridge, MA, 2007); and Gwendolyn Midlo Hall, *Slavery and African Ethnicities in the Americas: Restoring the Links* (Chapel Hill, NC, 2007).

24. See George Fredrickson, *White Supremacy: A Comparative Study of American and South African History* (New York, 1981); Peter Kolchin, *Unfree Labor: American Slavery and Russian*

Serfdom (Cambridge, MA, 1987); Shearer Davis Bowman, *Masters and Lords: Mid-Nineteenth Century U.S. Planters and Prussian Junkers* (New York, 1993); Enrico Dal Lago, *Agrarian Elites: American Slaveholders and Southern Italian Landowners, 1815–1861* (Baton Rouge, LA, 2005); Michael L. Bush, *Servitude in Modern Times* (Cambridge, 2000); and Dale Tomich, *Through the Prism of Slavery.* See also Immanuel Wallerstein, *The Modern World-System,* Vols. 3-4 (New York, 1989–2011).

25. Michael McGerr, "The Price of the "New Transnational History," *American Historical Review* 96 (1991), 1064; and Ian Tyrrell, *Transnational Nation: United States History in Global Perspective since 1789* (New York, 2007), 3.

26. David Thelen, "The Nation and Beyond: Transnational Perspectives on United States History," *Journal of American History* 86:3 (1999), 972; Thomas Bender, "Introduction: Historians, the Nation, and the Plenitude of Narratives" in Thomas Bender, ed., *Rethinking American History in a Global Age* (Berkeley, CA, 2002), 11-12. See also Jay Sexton, "The Global View of the United States," *Historical Journal* 48:1 (2005), 261-276.

27. Jurgen Kocka, "Comparison and Beyond," *History and Theory* 42 (2003), 43; Charles Bright and Michael Geyer, "Where in the World is America? The History of the United States in the Global Age" in Bender, ed., *Rethinking American History,* 64.

28. Heinz-Gerhard Haupt and Jurgen Kocka, "Comparative History: Methods, Aims, Problems" in Deborah Cohen and Maura O'Connor, eds., *Comparison and History: Europe in Cross-National Perspective* (London, 2004), 32-33; Ian Tyrrell, "Beyond the View from Euro-America: Environment, Settler Societies, and the Internationalization of American History" in Bender, ed., *Rethinking American History,* 169-171. See also Glenda Sluga, "The Nation and the Comparative Imagination" in Cohen and O'Connor, eds., *Comparison and History,* 103-114; and Heinz-Gerhard Haupt and Jurgen Kocka, eds., *Comparative History and the Quest for Transnationality* (Oxford, 2009). On *histoire croisée,* see Michael Werner and Benedicte Zimmermann, "Beyond Comparison: Histoire Croisée and the Challenge of Reflexivity," *History and Theory* 45 (2006), 30-50. On *transfergeschichte,* see Michel Espagne, *Les transfers culturels franco-allemands* (Paris, 1999). An important recent study that goes in the direction envisioned by Ian Tyrrell is James Belich, *Replenish the Earth: The Settler Revolution and the Rise of the Anglo-World, 1783–1939* (New York, 2009).

29. Michael Geyer and Charles Bright, "World History in a Global Age," *American Historical Review* 100:5 (1995), 140; Marcus Gräser, "World History in a Nation-State: The Transnational Disposition in Historical Writing in the United States," *Journal of American History* 95:4 (2009), 1040.

30. George Fredrickson, "From Exceptionalism to Variability: Recent Developments in Cross-National Comparative History," *Journal of American History* 82:2 (1995), 604.

31. George Fredrickson, "From Exceptionalism to Variability," 604.

32. Deborah Cohen and Maura O'Connor, "Introduction" in Cohen and O'Connor, eds., *Comparison and History,* xii-xiii.

33. Deborah Cohen and Maura O'Connor, "Introduction," xii-xiii; and Michael Miller, "Comparative and Cross-National History: Approaches, Differences, Problems," in Cohen and O'Connor, eds., *Comparison and History,* 116, 126-127.

34. On world history and comparative history of slavery in synchronic and diachronic perspectives, see Enrico Dal Lago and Constantina Katsari, "The Study of Ancient and Modern Slave Systems: Setting an Agenda for Comparison" in Dal Lago and Katsari, eds., *Slave Systems: Ancient and Modern* (New York, 2008), 3-31. See also Trevor Burnard and Gad Heuman, "Introduction" in Heuman and Burnard, eds., *The Routledge History of Slavery* (New York, 2011), 1-16.

35. C. A. Bayly, *The Birth of the Modern World, 1780–1914: Global Connections and Comparisons* (Oxford, 2004), 1. The recent Sinocentric approach is represented most notably by Kenneth Pomeranz, *The Great Divergence: China, Europe, and the Making of the Modern World Economy* (Princeton, NJ, 2000).

36. See, for a very recent example, Emma Rotschild, "Late Atlantic History" in Nicholas

Canny and Philip Morgan, eds., *The Oxford Handbook of the Atlantic World, 1450–1850* (New York, 2011), 634-648.

CHAPTER ONE

1. See William Phillips, *Slavery from Roman Times to the Early Transatlantic Trade* (New York, 1985).

2. Michael L. Bush, *Servitude in Modern Times* (Cambridge, 2000), 69. See also David Eltis and Stanley Engerman, "Dependence, Servility, and Coerced Labor in Time and Space" in Eltis and Engerman, eds., *The Cambridge World History of Slavery,* Vol. 3: *AD 1420–AD 1804* (New York, 2011), 1-23.

3. See David Eltis, *The Rise of African Slavery in the Americas* (New York, 1999), 258-280. On the simultaneous rise of slavery in the Americas and serfdom in eastern Europe, see Peter Kolchin, *Unfree Labor: American Slavery and Russian Serfdom* (Cambridge, MA, 1987); and Richard Ellie, "Russian Slavery and Serfdom, 1450–1804" in Eltis and Stanley Engerman, eds., *Cambridge World History of Slavery,* Vol. 3, 275-296.

4. See Robin Blackburn, *The Making of New World Slavery, 1492–1800: From the Baroque to the Modern* (London, 1997); and Linda Heywood and John Thornton, *Central Africans, Atlantic Creoles, and the Foundations of the Americas* (New York, 2007). On rice in particular, see Judith Carney, *Black Rice: The African Origins of Rice Cultivation in the Americas* (Cambridge, MA, 2001); and David Eltis et al., "Agency and Diaspora in Atlantic History: Reassessing the African Contribution to Rice Cultivation in the Americas," *American Historical Review* 112:5 (December 2007), 1329-1358.

5. See David Eltis and David Richardson, *Atlas of the Transatlantic Slave Trade* (New Haven, CT, 2010); and Frederick C. Knight, *Working the Diaspora: The Impact of African Labor on the Anglo-American World, 1650–1850* (New York, 2010).

6. See John Thornton, *Africa and Africans in the Making of the Atlantic World, 1400–1800* (New York, 1998), 72-98; and Victor Ehrenberg, *A History of Africa to 1800* (Charlottesville, VA, 1997).

7. See Paul Lovejoy, *Transformations in Slavery: A History of Slavery in Africa* (New York, 2000).

8. Charles Verlinden, *L'esclavage dans l'Europe medieval,* 2 vols. (Ghent, 1955-1977).

9. See Philip D. Curtin, *The Rise and Fall of the Plantation Complex: Essays in Atlantic History* (New York, 1990), 3-28. See also Sidney Mintz, *Sweetness and Power: The Place of Sugar in Modern History* (New York, 1985).

10. See Jason W. Moore, "Madeira, Sugar, and the Conquest of Nature in the 'First' Sixteenth Century, Part I: From 'Island of Timber' to Sugar Revolution, 1420–1506," *Review* 32:4 (2009), 345-390; Robin Blackburn, *Making of New World Slavery*; and David Brion Davis, *Inhuman Bondage: The Rise and Fall of Slavery in the New World* (New York, 2006), 103-123.

11. On the concept of "middle passages," see Marcus Rediker et al., "Introduction" in Emma Christopher et al., *Many Middle Passages: Forced Migration and the Making of the Modern World* (Berkeley, CA, 2007), 1-19. See also Herbert S. Klein, *The Atlantic Slave Trade* (New York, 2010), 74-102; Klein, *The Middle Passage: Comparative Studies in the Atlantic Slave Trade* (Princeton, NJ, 1978); and James Walvin, *The Slave Trade* (New York, 2011), 47-66.

12. See Luiz Felipe de Alencastro, "The Apprenticeship to Colonization" in Barbara L. Solow, ed., *Slavery and the Rise of the Atlantic System* (New York, 1991), 151-176.

13. Thomas Benjamin, *The Atlantic World: Europeans, Africans, Indians, and Their Shared History, 1400–1900* (New York, 2009), 344. See also Luiz Felipe de Alencastro, *O trato dos viventes: Formaçao do Brasil no Atlantico Sul, seculos XVI e XVII* (São Paulo, 2000).

14. The statistical data is in Thomas Benjamin, *The Atlantic World,* 342. See also M. D. D. Newitt, *A History of Portuguese Overseas Expansion, 1400–1668* (London, 2005).

15. See John H. Elliott, *Empires of the Atlantic World: Britain and Spain in America, 1492–1830* (New Haven, CT, 2006).

16. See Leslie Byrd Simpson, *The Encomienda in New Spain: The Beginnings of Spanish Mexico* (Berkeley, CA, 1966); and William D. Phillips, "Slavery in the Atlantic Islands and in the Early Modern Spanish Atlantic World" in David Eltis and Stanley Engerman, eds., *Cambridge World History of Slavery,* Vol. 3, 325-349.

17. Stuart Schwartz, "The Iberian Atlantic to 1650" in Nicholas Canny and Philip D. Morgan, eds., *The Oxford Handbook of the Atlantic World 1450–1850* (New York, 2011), 147-164.

18. See John H. Elliott, *Spain, Europe, and the Wider World, 1500–1800* (New Haven, CT, 2009).

19. David Brion Davis, *Inhuman Bondage,* 90. The statistical data is in Thomas Benjamin, *The Atlantic World,* 342. See also Kenneth J. Andrien, "The Spanish Atlantic System" in Jack P. Greene and Philip Morgan, eds., *Atlantic History: A Critical Appraisal* (New York, 2009), 55-79.

20. See Piet C. Emmer, *The Dutch Slave Trade, 1500–1850* (New York, 2006). See also Wim Klooster, "The Northern European Atlantic World" in Nicholas Canny and Philip D. Morgan, eds., *Oxford Handbook of the Atlantic World,* 165-182.

21. See David P. Geggus, "The French Slave Trade: An Overview," *William and Mary Quarterly* 58 (2001), 119-138.

22. The statistical data is in Thomas Benjamin, *The Atlantic World,* 342. See Silvia Marzagalli, "The French Atlantic World in the Seventeenth and Eighteenth Centuries" in Nicholas Canny and Philip D. Morgan, eds., *Oxford Handbook of the Atlantic World,* 235-251.

23. See Kenneth Morgan, *Slavery and the British Empire: From Africa to America* (New York, 2007), 7-33; and David Richardson, "The British Empire and the Atlantic Slave Trade, 1660–1800" in Peter J. Marshall, ed., *The Oxford History of the British Empire,* Vol. 7: *The Eighteenth Century* (Oxford, 1998). See also Joyce Chaplin, "The British Atlantic" in Nicholas Canny and Philip D. Morgan, eds., *Oxford Handbook of the Atlantic World,* 219-234.

24. The statistical data is from Trevor Burnard, "The Atlantic Slave Trade" in Gad Heuman and Burnard, eds., *The Routledge History of Slavery* (London, 2011), 91. See also Kenneth Morgan, *Slavery, Atlantic Trade, and the British Economy, 1660–1800* (Cambridge, 2000).

25. David Eltis, "The U.S. Transatlantic Slave Trade, 1644–1867: An Assessment," *Civil War History* 54 (2008), 357; and Herbert S. Klein, *Atlantic Slave Trade,* 201-211. See also Ira Berlin, *The Making of African America; The Four Great Migrations* (New York, 2010), 49-98.

26. See Colin A. Palmer, "The Middle Passage" in Beverly C. McMillan, ed., *Captive Passage: The Transatlantic Slave Trade and the Making of the Americas* (Washington, DC, 2002), 53-77; and Marcus Rediker, *The Slave Ship: A Human History* (London, 2007).

27. See James Walvin, *The Slave Trade,* 67-83.

28. For the distinction between "slave society" and "society with slaves," see Moses Finley, *Ancient Slavery and Modern Ideology* (London, 1980), 135-160.

29. On the concept of "many souths," see Peter Kolchin, *A Sphinx on the American Land: The Nineteenth-Century South in Comparative Perspective* (Baton Rouge, LA, 2003), 39-73. On "commodity frontiers," see Jason W. Moore, "Sugar and the Expansion of the Early Modern World-Economy: Commodity Frontiers, Ecological Transformation, and Industrialization," *Review* 23:3 (2000), 409-433.

30. See Stuart Schwartz, *Sugar Plantations in the Formation of Brazilian Society: Bahia, 1550–1835* (New York, 1985).

31. Stuart Schwartz, "A Commonwealth Within Itself: The Early Brazilian Sugar Industry, 1550–1670" in Schwartz, ed., *Tropical Babylons: Sugar and the Making of the Atlantic World, 1450–1680* (Chapel Hill, NC, 2004), 180.

32. See Frederic Mauro, "Political and Economic Structures of Empire, 1580–1750" in Leslie Bethell, ed., *Colonial Brazil* (New York, 1987), 39-66; and João Fragoso and Ana

Rio, "Slavery and Politics in Colonial Portuguese America; The Sixteenth to the Eighteenth Centuries" in David Eltis and Stanley Engerman, eds., *Cambridge World History of Slavery,* Vol. 3, 350-377.

33. See Herbert S. Klein and Francisco Vidal Luna, *Slavery in Brazil* (New York, 2010), 19-34.

34. See Barry W. Higman, "The Sugar Revolution," *Economic History Review* 53 (2000), 213-238.

35. See Richard S. Dunn, *Sugar and Slaves: The Rise of the Planter Class in the English West Indies, 1624–1713* (Chapel Hill, NC, 1972).

36. John McCusker and Russell R. Menard, "The Sugar Industry in the Seventeenth Century: A New Perspective on the Barbadian 'Sugar Revolution'" in Stuart Schwartz, ed., *Tropical Babylons,* 306.

37. See Russell R. Menard, *Sweet Negotiations: Sugar, Slavery, and Plantation Agriculture in Early Barbados* (Charlottesville, VA, 2006).

38. See Philip Morgan, ed., *Unity and Diversity in Early North America* (London, 1993).

39. See Daniel Littlefield, "Colonial and Revolutionary United States" in Robert L. Paquette and Mark M. Smith, eds., *The Oxford Handbook of Slavery in the Americas* (New York, 2010), 201-226; and April Lee Hatfield, *Atlantic Virginia: Intercolonial Relations in the Seventeenth Century* (Philadelphia, PA, 2004).

40. See Edmund Morgan, *American Slavery, American Freedom: The Ordeal of Colonial Virginia* (New York, 1975).

41. See Ira Berlin, *Many Thousands Gone: the First Two Centuries of Slavery in North America* (Cambridge, MA, 1998), 29-76; and Alden T. Vaughan, *Roots of American Racism: Essays on the Colonial Debate* (New York, 1995).

42. See Jack P. Greene, "Colonial South Carolina and the Caribbean Connection," *South Carolina Historical Magazine* 88 (1987), 192-210; and S. Max Edelson, *Plantation Enterprise in Colonial South Carolina* (Cambridge, MA, 2006).

43. See S. Max Edelson, "Beyond Black Rice: Reconstructing Material and Cultural Context for Early Plantation Agriculture," *American Historical Review* 115:1 (2010), 125-135.

44. See Peter Coclanis, *The Shadow of a Dream: Economic Life and Death in the South Carolina Low Country,* 1660–1920 (New York, 1989).

45. See Robin Blackburn, *The American Crucible: Slavery, Emancipation, and Human Rights* (London, 2011), 49-76.

46. See Philip D. Morgan, *Slave Counterpoint: Black Eighteenth-Century Culture in the Chesapeake and Lowcountry* (Chapel Hill, NC, 1998), 1-23.

47. See Gwendolyn Midlo Hall, *Slavery and African Ethnicities in the Americas: Restoring the Links* (Chapel Hill, NC, 2005), 49-50, 68-69; and Sidney Mintz and Richard Price, *The Birth of African-American Culture: An Anthropological Perspective* (Boston, 1992, orig. pub. in 1976).

48. On African-led slave rebellions in the Americas, see Gad Heuman, "Slave Rebellions" in Heuman and Trevor Burnard, eds., *Routledge History of Slavery,* 220-233.

49. See Linda Heywood and John Thornton, *Central Africans, Atlantic Creoles,* 109-168.

50. See David Northrup, "Africans, Early European Contacts, and the Emergent Diaspora" in Nicholas Canny and Philip Morgan, eds., *Oxford Handbook of the Atlantic World,* 38-54.

51. João José Reis and Flavio dos Santos Gomez, eds., *Libertade por um fio. Historia dos Quilombos no Brazil* (São Paulo, 1996).

52. See Mary Karasch, "Zumbi of Palmares: Challenging the Portuguese Colonial Order" in Kenneth J. Andrien, ed., *The Human Tradition in Colonial Latin America* (Wilmington, DE, 2002), 104-120.

53. Stuart Schwartz, *Slaves, Peasants, and Rebels: Reconsidering Brazilian Slavery* (Madison, WI, 1992), 125.

54. See Philip D. Morgan, "Slavery in the British Caribbean" in David Eltis and Stanley Engerman, eds., *Cambridge World History of Slavery,* Vol. 3, 378-406.

55. David Eltis, *Rise of African Slavery,* 255; Trevor Burnard, "British West Indies and

Bermuda" in Robert L. Paquette and Mark M. Smith, eds., *Oxford Handbook of Slavery in the Americas,* 138.

56. See David Barry Gaspar, "With a Rod of Iron: Barbados Slave Laws as a Model for Jamaica, South Carolina, and Antigua, 1661-1697" in Darlene Clark Hine and Jacqueline McLeod, eds., *Comparative History of Black People in Diaspora* (Bloomington, IN, 1999), 343-366.

57. See Morgan, *Slavery and the British Empire,* 111-114. See also Hilary M. Beckles, *White Servitude and Black Slavery in Barbados, 1627–1715* (Knoxville, TN, 1989).

58. John Rolfe, "The First Blacks Arrive in Virginia (1619)" in Rick Halpern and Enrico Dal Lago, eds., *Slavery and Emancipation* (Oxford, 2002), 13. See also Midlo Hall, *Slavery and African Ethnicities,* 90-91; and Michael A. Gomez, *Exchanging Our Country Marks: The Transformation of African Identities in the Colonial and Antebellum South* (Chapel Hill, NC, 1998).

59. See Lorena S. Walsh, *From Calabar to Castle Grove: The History of a Virginia Slave Community* (Charlottesville, VA, 2001), 53-81.

60. See Alan Kulikoff, *Tobacco and Slaves: The Development of Southern Cultures in the Chesapeake, 1660–1800* (Chapel Hill, NC, 1986).

61. See Anthony Parent, *Foul Means: The Formation of Slave Society in Virginia, 1660–1740* (Chapel Hill, NC, 2003), 105-134.

62. See Daniel C. Littlefield, *Rice and Slaves: Ethnicity and the Slave Trade in Colonial South Carolina* (Urbana, IL, 1991), 8-32.

63. See Betty Wood, *Slavery in Colonial America, 1619–1776* (New York, 2005), 60-69.

64. See Peter H. Wood, *Black Majority: Negroes in Colonial South Carolina from 1670 through the Stono Rebellion* (New York, 1974).

65. See Peter Charles Hoffer, *Cry Liberty: The Great Stono Slave Rebellion of 1739* (New York, 2010).

66. See John K. Thornton, "African Dimensions" in Mark M. Smith, *Stono: Documenting and Interpreting a Southern Slave Revolt* (Columbia, SC, 2005), 73-86.

67. Ira Berlin, *Many Thousands Gone,* 12.

68. Ira Berlin, *Many Thousands Gone,* 17. See also Betty Wood, "The Origins of Slavery in the Americas, 1500–1700" in Gad Heuman and Trevor Burnard, *Routledge History of Slavery,* 64-79.

69. See Immanuel Wallerstein, *The Modern World-System,* Vols. 1–2 (New York, 1974–1981); and Eric Wolf, *Europe and the People without History* (Berkeley, CA, 1982).

70. See Nicholas Canny, ed., *The Oxford History of the British Empire,* Vol. 1: *The Origins of Empire: British Overseas Enterprise to the Close of the Seventeenth Century* (Oxford, 1998); and David Armitage and Michael Braddick, eds., *The British Atlantic World, 1500–1800* (New York, 2002).

71. Robin Blackburn, *Origins of New World Slavery*; and Rafael de Bivar Marquese, *Feitores do corpo, missionarios da mente: Senhores, letrados e o controle dos escravos nas Americas, 1660–1860* (São Paulo, 2004).

72. See Stephanie Smallwood, *Saltwater Slavery: A Middle Passage from Africa to America Diaspora* (Cambridge, MA, 2007); and Paul Gilroy, *The Black Atlantic: Modernity and Double Consciousness* (Cambridge, MA, 1993).

73. See Linda Heywood and John Thornton, *Central Africans, Atlantic Creoles*; Gwendolyn Midlo Hall, *Slavery and African Ethnicities in the Americas*; and Philip D. Morgan, "British Encounters with Africans and African-Americans, circa 1600–1780" in Bernard Bailyn and Philip Morgan, eds., *Strangers within the Realm: Cultural Margins in the First British Empire* (Chapel Hill, NC, 1991), 157-219.

74. See Immanuel Wallerstein, *Modern World System,* Vol. 1; Witold Kula, *An Economic Study of the Feudal System* (London, 1976); and Edgar Melton, "Manorialism and Rural Subjection in East Central Europe, 1500–1800" in David Eltis and Stanley Engerman, eds., *Cambridge World History of Slavery,* Vol. 3, 297-323.

75. See Eric Williams, *Capitalism and Slavery* (Chapel Hill, NC, 1944); and Seymour Drescher, *From Slavery to Freedom: Comparative Studies in the Rise and Fall of Atlantic Slavery* (New York, 1999).

CHAPTER TWO

1. See Robin Blackburn, *The Making of New World Slavery: From the Baroque to the Modern, 1492–1800* (London, 1997), 371-580.

2. On cultural transfers in the late eighteenth-century Americas' slave societies, see Michael Zeuske, "Comparing or Interlinking? Economic Comparisons of Early Nineteenth-Century Slave Systems in the Americas in Historical Perspective" in Enrico Dal Lago and Constantina Katsari, eds., *Slave Systems: Ancient and Modern* (New York, 2008), 148-183.

3. See Laurent Dubois, "Slavery in the Age of Revolutions" in Gad Heuman and Trevor Burnard, eds., *The Routledge History of Slavery* (London, 2011), 267-280.

4. See Ira Berlin, *Many Thousands Gone: The First Two Centuries of Slavery in North America* (Cambridge, MA, 1998), 97-98. See also Jason W. Moore, "Sugar and the Expansion of the Early Modern World-Economy: Commodity Frontiers, Ecological Transformation, and Industrialization," *Review* 23:3 (2000), 409-433.

5. On the definition of slave society, see Moses Finley, *Ancient Slavery and Modern Ideology* (London, 1980), 135-160; Keith Hopkins, *Conquerors and Slaves* (New York, 1978), 99-100; and Ira Berlin, *Many Thousands Gone*, 10-11.

6. See David Eltis, "Africa, Slavery, and the Slave Trade, mid-Seventeenth Century to mid-Eighteenth Century" in Nicholas Canny and Philip D. Morgan, eds., *The Oxford Handbook of the Atlantic World, 1450–1850* (New York, 2011), 271-288.

7. See Eugene Genovese, "The Slave Systems and Their European Antecedents" in *The World the Slaveholders Made: Two Essays in Interpretation* (New York, 1968), 21-101. See also Peter Kolchin, *American Slavery, 1619–1877* (New York, 2003), 28-62.

8. See Peter Kolchin, *American Slavery*, 33-34.

9. On conspicuous consumption, see Thornstein Veblen, *The Theory of the Leisure Class* (New York, 1899). See also, on Virginia, Lorena S. Walsh, *Motives of Honor, Pleasure, and Profit: Plantation Management in the Colonial Chesapeake, 1607–1763* (Chapel Hill, NC, 2010).

10. On South Carolina, see Robert Olwell, *Masters, Slaves, and Subjects: The Culture of Power in the South Carolina Low Country, 1740–1790* (Ithaca, NY, 1998).

11. See Lorena S. Walsh, "Slavery in the North American Mainland Colonies" in David Eltis and Stanley Engerman, eds., *The Cambridge World History of Slavery*, Vol. 3: *AD 1420–AD 1804* (New York, 2011), 407-430.

12. On task system and gang system, see Ira Berlin and Philip D. Morgan, "Labor and the Shaping of Slave Life in the Americas" in Berlin and Morgan, eds., *Cultivation and Culture: Labor and the Shaping of Slave Life in the Americas* (Charlottesville, VA, 1993), 1-48.

13. John D. Blassingame, *The Slave Community: Plantation Life in the Antebellum South* (New York, 1979), 49. See also Philip D. Morgan, *Slave Counterpoint: Black Life in the Eighteenth-Century Chesapeake and the Lowcountry* (Chapel Hill, NC, 1998). For the statistical data, see Richard H. Steckel, "Demography and Slavery" in Robert L. Paquette and Mark M. Smith, eds., *The Oxford Handbook of Slavery in the Americas* (New York, 2010), 650-651.

14. See Michael A. Gomez, *Exchanging Our Country Marks: The Transformation of African Identities in the Colonial and Antebellum South* (Chapel Hill, NC, 1998), 244-291; and James Sidbury, *Becoming African in America: Race and Nation in the Early Black Atlantic* (New York, 2007).

15. See also for the statistics Richard B. Sheridan, *Sugar and Slavery: An Economic History of the British West Indies, 1623–1775* (Barbados, 1974), 208-233.

16. Gad Heuman, *The Caribbean* (New York, 2006), 46. See also B. W. Higman, *Plantation Jamaica, 1750–1850: Capital and Control in a Colonial Economy* (Kingston, 2005).

17. See Trevor Burnard, *Mastery, Tyranny, and Desire: Thomas Thistlewood and His Slaves in*

the Anglo-Jamaican World (Chapel Hill, NC, 2004); and Vincent Brown, *The Reaper's Garden: Death and Power in the World of Atlantic Slavery* (Cambridge, MA, 2008).

18. See Kamau Brathwaite, *The Development of Creole Society in Jamaica, 1770–1820* (Oxford, 1971), 215-230.

19. Philip D. Morgan, "The Black Experience in the British Empire, 1680–1810" in Peter J. Marshall, ed., *The Oxford History of the British Empire,* Vol. II: *The Eighteenth Century* (New York, 1998), 482.

20. Stuart Schwartz, "Plantations and Peripheries, c. 1580—c. 1750" in Leslie Bethell, ed., *Colonial Brazil* (New York, 1987), 89.

21. See Stuart Schwartz, *Sugar Plantations in the Formation of Brazilian Society: Bahia, 1550–1835* (New York, 1985).

22. See Herbert S. Klein and Francisco Vidal Luna, *Slavery in Brazil* (New York, 2010), 35-52.

23. See Laird Bergad, *Slavery and the Demographic and Economic History of Minas Gerais, 1720–1888* (New York, 1999); and Francisco Vidal Luna, *Minas Gerais: Escravos e senhores* (São Paulo, 1980).

24. See Herbert Klein and Francisco Vidal Luna, *Slavery in Brazil,* 53-55.

25. See Elizabeth W. Kiddy, *Blacks of the Rosary: Memory and History in Minas Gerais* (University Park, PA, 2005); and Celia Maria Borges, *Escravos e libertos nas Irmandades do Rosario: Devoçao e solidaridade em Minas Gerais, seculos XVIII e XIX* (Juiz de Fora, Minas Gerais, 2005).

26. See Ira Berlin, *Many Thousands Gone,* 12-13; Gwendolyn Midlo Hall, *Slavery and African Ethnicities in the Americas: Restoring the Links* (Chapel Hill, NC, 2005); and Matt D. Childs, "Slave Culture" in Gad Heuman and Trevor Burnard, eds., *Routledge History of Slavery,* 170-186.

27. See Trevor Burnard, "The Planter Class" in Gad Heuman and Burnard, eds., *Routledge History of Slavery,* 187-203.

28. See Eugene D. Genovese and Douglas Ambrose, "Masters" in Robert L. Paquette and Mark M. Smith, eds., *The Oxford Handbook of Slavery in the Americas,* 535-555.

29. See T. H. Breen, *Tobacco Culture: the Mentality of the Great Tidewater Planters on the Eve of Revolution* (Princeton, NJ, 2001), 3-39.

30. See Rhys Isaac, *The Transformation of Virginia, 1740–1790* (Chapel Hill, NC, 1982), 18-42, 235-262.

31. See Enrico Dal Lago, "Patriarchs and Republicans: Eighteenth-Century Virginian Planters and Classical Politics," *Historical Research* 76 (2003), 492-511.

32. See Rhys Isaac, *Landon Carter's Uneasy Kingdom: Revolution and Rebellion on a Virginia Plantation* (New York, 2005), 57-122; and Joyce Chaplin, *An Anxious Pursuit: Agricultural Innovation and Modernity in the Lower South, 1730–1815* (Chapel Hill, NC, 1993).

33. See Charles F, Irons, *The Origins of Proslavery Christianity: White and Black Evangelicals in Colonial and Antebellum Virginia* (Chapel Hill, NC, 2008).

34. William Byrd II's quote is in Rhys Isaac, *Transformation of Virginia,* 40. See also Willie Lee Rose, "The Domestication of Domestic Slavery" in William W. Freehling, ed., *Slavery and Freedom* (New York, 1982), 18-36.

35. See James Walvin, *Black Ivory: A History of British Slavery* (London, 2001), 3-23; and Trevor Burnard, "Powerless Masters: The Curious Decline of Jamaican Sugar Planters in the Foundational Period of British Abolitionism," *Slavery & Abolition* 32:2 (2011), 185-298.

36. See David Brion Davis, *Slavery and Human Progress* (New York, 1984), 107-128; and Robin Blackburn, *The Overthrow of Colonial Slavery, 1776–1848* (London, 1988), 67-109, 131-160.

37. See J. R. Ward, *British West Indian Slavery, 1750–1834: The Process of Amelioration* (Oxford, 1988).

38. J. R. Ward, "The British West Indies in the Age of Abolition, 1748–1815" in Peter J. Marshall, ed., *The Eighteenth Century,* 429-430. See also David Beck Ryden, *West Indian*

Slavery and British Abolition, 1783–1807 (New York, 2009); Christopher Brown, *Moral Capital: Foundations of British Abolitionism* (Chapel Hill, NC, 2006); and Philip D. Morgan, "Ending the Slave Trade: A Caribbean and Atlantic Context" in Derek R. Peterson, ed., *Abolitionism and Imperialism in Britain, Africa, and the Atlantic* (Athens, OH, 2010), 101-128.

39. See John D. Garrigus, "The French Caribbean" in Robert L. Paquette and Mark M. Smith, eds., *Oxford Handbook of Slavery in the Americas,* 173-200.

40. See Laurent Dubois and John D. Garrigus, "Introduction: Revolution, Emancipation, and Independence" in Dubois and Gerrigus, eds., *Slave Revolution in the Caribbean, 1789–1804* (New York, 2006), 7-46.

41. See David Geggus, "Sugar and Coffee Cultivation in Saint Domingue and the Shaping of the Slave Labor Force" in Ira Berlin and Philip D. Morgan, eds., *Cultivation and Culture,* 73-100; and Gabriel Debien, *Les Esclaves aux Antilles francaises, XVIIe-XVIIIe siecles* (Basse-Terre, 1974).

42. Robin Blackburn, *Overthrow of Colonial Slavery,* 44. See also John Garrigus, *Before Haiti: Race and Citizenship in French Saint-Domingue* (New York, 2006); and Stewart King, *Blue Coat or Powdered Wig: Free People of Color in Pre-Revolutionary Saint-Domingue* (Athens, GA, 2000).

43. See Laurent Dubois, "Slavery in the French Caribbean, 1635–1804" in David Eltis and Stanley Engerman, eds., *Cambridge World History of Slavery,* Vol. 3, 431-449.

44. See Robin Blackburn, *The American Crucible: Slavery, Emancipation, and Human Rights* (London, 2011), esp. 171-274.

45. See Arthur L. Stinchcombe, *Sugar Island Slavery in the Age of Enlightenment: The Political Economy of the Caribbean World* (Princeton, NJ, 1995).

46. See Christopher Brown, "The Abolition of the Slave Trade" in Gad Heuman and Trevor Burnard, eds., *Routledge History of Slavery,* 281-297; and Laurent Dubois, "Slavery in the Age of Revolution," 269-276.

47. See David Brion Davis, *The Problem of Slavery in the Age of Revolutions, 1770–1823* (Ithaca, NY, 1975); and Eugene D. Genovese, *From Rebellion to Revolution: Afro-American Slave Revolts in the Making of the Modern World* (Baton Rouge, LA, 1979).

48. See Robin Blackburn, *Overthrow of Colonial Slavery*; and Blackburn, *American Crucible.* See also David Armitage and Sanjay Subrahnmanyam, "The Age of Revolutions, c. 1760–1840—Global Causation, Connection, and Comparison" in Armitage and Subrahnmanyam, eds., *The Age of Revolutions in Global Context, c. 1760-1840* (New York, 2010), xii-xxxii.

49. See Robin Blackburn, "Haiti, Slavery, and the Age of Democratic Revolutions," *William & Mary Quarterly* 63 (2006), 643-674.

50. See David Brion Davis, "Impact of the French and Haitian Revolutions" in David P. Geggus, ed., *The Impact of the Haitian Revolution on the Atlantic World* (Columbia, SC, 2001), 3-9; and Geggus, "The Haitian Revolution in Atlantic Perspective" in Nicholas Canny and Philip D. Morgan, eds., *Oxford Handbook of the Atlantic World,* 533-549; and Manolo Florentino and Marcia Amantino, "Runaways and *Quilombolas* in the Americas" in David Eltis and Stanley Engerman, eds., *Cambridge World History of Slavery,* Vol. 3, 708-740.

51. See Peter Kolchin, *American Slavery,* 63-92; and Duncan J. MacLeod, *Slavery, Race, and the American Revolution* (New York, 1974).

52. See Philip Morgan, *Slave Counterpoint,* 666-667. See also Cassandra Pybus, *Epic Journeys: Runaway Slaves of the American Revolution and Their Global Quest for Liberty* (Boston, 2007).

53. See Ira Berlin, *Many Thousands Gone,* 256-324; ; and Daniel Littlefield, *Revolutionary Citizens: African Americans, 1776–1804* (New York, 1997).

54. See Ira Berlin, *Generations of Captivity: A History of African-American Slaves* (Cambridge, MA, 2003), 102-111.

55. See Don E. Fehrenbacher, *The Slaveholding Republic: An Account of the United States Government's Relations to Slavery* (New York, 2001), 15-48. See also Paul Finkelman, *Slavery*

and the Founders: Race and Liberty in the Age of Jefferson (Armonk, NY, 2001); and Gary B. Nash, *The Forgotten Fifth: African Americans in the Age of Revolution* (Cambridge, MA, 2006).

56. See Michael Craton, *Testing the Chains: Resistance to Slavery in the British West Indies* (Ithaca, NY, 1982).

57. Philip D. Morgan, "The Black Experience," 484. See also Mavis C. Campbell, *The Maroons of Jamaica, 1655–1796: A History of Resistance, Collaboration, and Betrayal* (Trenton, NJ, 1990).

58. See Gad Heuman, *The Caribbean,* 55-66; and Philip D. Morgan, "Slavery in the British Caribbean" in David Eltis and Stanley Engerman, eds., *Cambridge World History of Slavery,* Vol. 3, 393-400.

59. Herbert Klein and Francisco Vidal Luna, *Slavery in Brazil,* 197. See also Carlos Magno Guimaraes, "Mineraçao, quilombos e Palmares: Minas Gerais no seculo XVIII" in João José Reis and Flavio dos Santos Gomes, eds., *Libertade por um fio. Historia dos Quilombos no Brasil* (Sao Paulo, 1996).

60. Dauril Alden, "Late Colonial Brazil, 1750–1808" in Bethell, ed., *Colonial Brazil,* 337.; See also Hendrik Kraay, *Race, State, and Armed Forces in Independence-Era Brazil: Bahia, 1790s–1840s* (Stanford, CA, 2001); and Leslie Bethell, "The Independence of Brazil" in Bethell, ed., *Brazil: Empire and Republic, 1822–1930* (New York, 1989), 3-45.

61. Robin Blackburn, *American Crucible,* 206; and Michele Hector and Laennec Hurbon, "Introduction" in Hector and Hurbon, eds., *Genese de l'etat haitien (1804–1859)* (Port-au-Prince, 2008), 16-17.

62. C. L. R. James, *The Black Jacobins: Toussaint L'Ouverture and the San Domingo Revolution* (New York, 1963, orig. pub. in 1938); and Eugene Genovese, *From Rebellion to Revolution.* See also David Geggus, "The Caribbean in the Age of Revolution" in David Armitage and Sanjay Subrahnmanyam, eds., *The Age of Revolutions in Global Context,* 83-100; and the interesting comparison with Guadalupe in Laurent Dubois, *A Colony of Citizens: Revolution and Slave Emancipation in the French Caribbean, 1787–1804* (Chapel Hill, NC, 2004).

63. See Carolyn Fick, "Revolutionary Saint Domingue and the Emerging Atlantic: Paradigms of Sovereignty," *Review* 31:2 (2008),121-144.

64. See Robin Blackburn, *American Crucible,* 173-175. See also Jeremy Popkin, *You Are All Free: The Haitian Revolution and the Abolition of Slavery* (New York, 2011); David Geggus, "Marronage, Voodoo, and the Saint-Domingue Slave Revolution of 1791" in Patricia Galloway and Philip P. Boucher, eds., *Proceedings of the Fifteenth Meeting of the French Colonial Historical Society* (Lanham, MD, 1992); and John K. Thornton, "I Am the Subject of the King of Kongo: African Ideology in the Haitian Revolution," *Journal of World History* 4 (1993), 181-214.

65. See Robin Blackburn, *American Crucible,* 182-196. See also Stewart R. King, "Slavery and the Haitian Revolution" in Robert L. Paquette and Mark M. Smith, eds., *Oxford Handbook of Slavery in the Americas,* 598-624.

66. See Laurent Dubois, *Avengers of the New World: The Story of the Haitian Revolution* (Cambridge, MA, 2004), 209-280; and Carolyn Fick, *The Making of Haiti: The Saint Domingue Revolution from Below* (Knoxville, TN, 1990).

67. See Jane Landers, *Atlantic Creoles in the Age of Revolutions* (Cambridge, MA, 2010); David Geggus, "The Haitian Revolution"; and David Barry Gaspar and Geggus, eds., *A Turbulent Time: The French Revolution and the Greater Caribbean* (Bloomington, IN, 1997).

68. See Michael Zeuske, "Comparing or Interlinking?" 148-183.

69. See Wim Klooster, *Revolutions in the Atlantic World: A Comparative History* (New York, 2009), 158-174. See also Christopher Brown, "Slavery and Antislavery, 1760–1820" in Nicholas Canny and Philip D. Morgan, eds., *Oxford Handbook of the Atlantic World,* 602-617; Gordon S. Brown, *Toussaint's Clause: The Founding Fathers and the Haitian Revolution* (Jackson, MS, 2005); and Ashli White, *Encountering Revolution: Haiti and the Making of the Early Republic* (Baltimore, MD, 2010).

70. See Robin Blackburn, *American Crucible,* 99-120; and James Belich, *Replenish the*

Earth: The Settler Revolution and the Rise of the Anglo-World, 1783–1939 (New York, 2009), 21-218.

71. See Dale Tomich and Michael Zeuske, "Introduction. The Second Slavery: Mass Slavery, World-Economy, and Comparative Microhistories," *Review* 31:2 (2008), 91-100. See also Dale Tomich, *Through the Prism of Slavery: Labor, Capital, and World Economy* (Lanham, MD, 2003), 56-74.

72. See C. A. Bayly, "The Age of Revolutions in Global Context: An Afterword" in David Armitage and Sanjay Subrahnmanyam, eds., *The Age of Revolutions in Global Context,* 299-217.

73. See Sue Peabody and Keila Grinberg, "Introduction: Slavery, Freedom, and the Law" in Peabody and Grinberg, eds., *Slavery, Freedom, and the Law in the Atlantic World* (New York, 2007), 1-28.

CHAPTER THREE

1. Dale Tomich and Michael Zeuske, "Introduction, The Second Slavery: Mass Slavery, World-Economy, and Comparative Microhistories," *Review* 31:2 (2008), 91; Jason W. Moore, "Sugar and the Expansion of the Early Modern World-Economy: Commodity Frontiers, Ecological Transformation, and Industrialization," *Review* 22:3 (2000), 410.

2. Jason Moore, "Sugar and the Expansion of the Early Modern World-Economy," 412. See also Immanuel Wallerstein, *The Modern World-System*, 4 vols. (New York, 1974-2011).

3. Dale Tomich and Michael Zeuske, "Introduction," 95, on the Atlantic plantation zone.

4. See Dale Tomich, "Atlantic History and World Economy: Concepts and Constructions," *Protosociology* 20 (2004), 102-121; and Immanuel Wallerstein, *The Modern World-System*, Vol. 3: *The Second Era of Great Expansion of the Capitalist World-Economy* (New York, 1989).

5. Dale Tomich and Michael Zeuske, "Introduction," 91. See also Tomich, *Through the Prism of Slavery: Labor, Capital, and World Economy* (Lanham, MD, 2004), specifically 56-71, on the second slavery.

6. Paul Lovejoy, *Slavery, Commerce, and Production in the Sokoto Caliphate of West Africa* (Trenton, NJ, 2005), 3. For the demographic data on the Americas, see Laird W. Bergad, *The Comparative Histories of Slavery in Brazil, Cuba, and the United States* (New York, 2007), 1-32.

7. On all these points, see Paul Lovejoy, *Slavery, Commerce, and Production in the Sokoto Caliphate.*

8. Michael Zeuske, "Comparing or Interlinking? Economic Comparisons of Early Nineteenth-Century Slave Systems in the Americas in Historical Perspective" in Enrico Dal Lago and Constantina Katsari, eds., *Slave Systems: Ancient and Modern* (New York, 2008), 174. See also, for comparisons, Laird Bergad, *Comparative Histories*; and Eric Wolf, *Europe and the People Without History* (Berkeley, CA, 1982).

9. For the definition of large-scale slave societies, see B. W. Highman, "Slave Societies" in Paul Finkelman and Joseph Miller, eds., *Macmillan Encyclopedia of World Slavery* (New York, 1997), 826-827.

10. See Paul Lovejoy, *Transformations in Slavery: A History of Slavery in Africa* (New York, 2000), 201-208; and Murray Last, *The Sokoto Caliphate* (London, 1967).

11. See especially D. W. Meinig, *The Shaping of America: A Geographical Perspective on 500 Year of History*, Vol. 2: *Continental America, 1800-1867* (New Haven, CT, 1995), 273-295.

12. On the process of expansion of cotton production, see Adam Rothman, *Slave Country: American Expansion and the Origins of the Deep South* (Cambridge, MA, 2005), 123-173; and Harry L. Watson, "Slavery and Development in a Dual Economy: The South and the Market Revolution" in Melvyn Stokes and Stephen Conway, eds., *The Market Revolution in America: Social, Political, and Religious Expressions, 1800–1880* (Charlottesville, VA, 1996), 44-45.

13. See Dale Tomich, "The Wealth of Empire: Francisco Arango y Parreño, Political Economy, and the Second Slavery in Cuba," *Comparative Studies in Society and History* 45

(2003), 4-28; and Reinaldo Funes Monzotes, *From Rainforest to Cane Field in Cuba: An Environmental History since 1492* (Chapel Hill, NC, 2008), 127-134.

14. See Leslie Bethell and José Murilo de Carvalho, "1822–1850" in Leslie Bethell, ed., *Brazil: Empire and Republic, 1822–1930* (New York, 1989), 89-94; and Rafael de Bivar Marquese, "African Diaspora, Slavery, and the Paraiba Valley Plantation Landscape: Nineteenth-Century Brazil," *Review* 31:2 (2008), 195-216.

15. See John Michael Vlach, *Back of the Big House: The Architecture of Plantation Slavery* (Chapel Hill, NC, 1993), 1-32.

16. See William K. Scarborough, *Masters of the Big House: Elite Slaveholders of the Mid-Nineteenth-Century South* (Baton Rouge, LA, 2003), 122-174.

17. See Manuel Moreno Fraginals, *The Sugarmill: The Socio-Economic Complex of Sugar in Cuba, 1760–1860* (New York, 1976), 151-152.

18. See Manuel Moreno Fraginals, *The Sugarmill*, 151-152.

19. On plantations in the Paraiba Valley, see especially Ricardo Salles, *E o Vale era o escravo: Vassouras, seculo XIX. Senhores e escravos no coraçao do Imperio* (Rio de Janeiro, 2008).

20. See Bryan Daniel McCann, "The Whip and the Watch: Overseers in the Paraiba Valley, Brazil," *Slavery & Abolition* 18:2 (1997), 30-47.

21. Solomon Northrup's quotation is in Willie Lee Rose, ed., *A Documentary History of Slavery in North America* (Athens, GA, 1976), 314. See also Mark M. Smith, "Old South Time in Comparative Perspective," *American Historical Review* 101:5 (1996), 1432-1469; and, on slave resistance, Eugene Genovese, *Roll, Jordan, Roll: The World the Slaves Made* (New York, 1974), especially 285-324.

22. Richard Follett, *The Sugar Masters: Planters and Slaves in Louisiana's Cane World, 1820–1860* (Baton Rouge, LA, 2005), 95. See also Ira Berlin, *Generations of Captivity: A History of African American Slaves* (Cambridge, MA, 2003), 77-78, 177-178; and Mark M. Smith, *Debating Slavery: Economy and Society in the Antebellum American South* (New York, 1998), 42-43.

23. See Ricardo Salles, *E o Vale era o escravo* on the Paraiba valley.

24. Rafael de Bivar Marquese, "African Diaspora, Slavery, and the Paraiba Valley," 202. See also Rafael de Bivar Marquese, *Administração e escravidão. Ideias sobre a gestão da agricultura escravista brasileira* (São Paulo, 1999).

25. See Lisa Yun, "Chinese Coolies and African Slaves in Cuba, 1847–74," *Journal of Asian American Studies* 4:2 (2001), 99-122.

26. Robert L. Paquette, "Cuba" in Paul Finkelman and Joseph Miller, eds., *Macmillan Encyclopedia of World Slavery*, 231. See also Rafael de Bivar Marquese, *Feitores do corpo, missionarios da mente. Senhores, letrados e o controle dos escravos nas Americas, 1660–1860* (São Paulo, 2004), 335-336.

27. See Paul Lovejoy, *Transformations in Slavery*, 201-208.

28. See Jennifer Lofkranz, "Protecting Freeborn Muslims: The Sokoto Caliphate's Attempts to Prevent Illegal Enslavement and Its Acceptance of the Strategy of Ransoming," *Slavery & Abolition* 32:1 (2011), 109-127; and H. A. S. Johnston, *The Fulani Empire of Sokoto* (London, 1967).

29. See Mohammed Bashir Salau, *The West African Slave Plantation: A Case Study* (New York, 2011).

30. Paul Lovejoy, "Plantations: Sokoto Caliphate and Western Sudan" in Paul Finkelman and Joseph Miller, eds., *Macmillan Encyclopedia of World Slavery*. Imam Imoru's citation is in Lovejoy, *Transformations in Slavery*, 205.

31. See Paul Lovejoy, "Plantations in the Economy of the Sokoto Caliphate," *Journal of African History* 19:3 (1978), 341-368.

32. On long-distance trade, especially in palm oil, in West Africa, see Martin Lynn, *Commerce and Economic Change in West Africa: The Palm Oil Trade in the Nineteenth Century* (New York, 1997).

33. See Paul Lovejoy, "The Characteristics of Plantations in Nineteenth-Century Sokoto Caliphate," *American Historical Review* 84:4 (1979), 1282-1283.

34. Anthony E. Kaye, "The Second Slavery: Modernity in the Nineteenth-Century South and the Atlantic World," *Journal of Southern History* 75:3 (2009), 634.

35. Michael Zeuske, "Comparing or Interlinking?" 165. See also Daniel Brett Rood, "Plantation Technocrats: A Social History of Knowledge in the Slaveholding Atlantic World, 1830–1865," Ph.D. Dissertation, University of California, Irvine, 2010.

36. See Paul Lovejoy and Stephen Baier, "The Desert-Side Economy of the Central Sudan," *International Journal of African Historical Studies* 8:4 (1975), 551-581.

37. See M. Last, "The Sokoto Caliphate and Borno" in J. F. Ade Ajayi, ed., *UNESCO General History of Africa*, Vol. VI: *Africa in the Nineteenth Century until 1880* (Berkeley, CA, 1998), 225-238.

38. See Adam Rothman, *Slave Country*, 37-119; James David Miller, *South by Southwest: Planter Emigration and Identity in the Slave South* (Charlottesville, VA, 2002); Edward Baptist, *Creating an Old South: Middle Florida's Plantation Frontier before the Civil War* (Chapel Hill, NC, 2002); and Richard Follett, *The Sugar Masters*, 14-45.

39. See Mark M. Smith, "Old South Time in Comparative Perspective," 1432-1469; Robert Gudmestad, *Steamboats and the Rise of the Cotton Kingdom* (Baton Rouge, LA, 2011); and Anthony Kaye, "The Second Slavery", 627-650.

40. Ira Berlin, *Generations of Captivity*, 166-169. See also Robert H. Gudmestad, *A Troublesome Commerce: The Transformation of the Interstate Slave Trade* (Baton Rouge, LA, 2003); and Walter Johnson, *Soul by Soul: Life Inside the Antebellum Slave Market* (Cambridge, MA, 1999).

41. See Steven G. Collins, "System, Organization, and Agricultural Reform in the Antebellum South, 1840–1860," *Agricultural History* 75 (2001), 1-27; and Enrico Dal Lago, *Agrarian Elites: American Slaveholders and Southern Italian Landowners, 1815–1861* (Baton Rouge, LA, 2005), 82-84, 294-298.

42. See Laurence Shore, *Southern Capitalists: The Ideological Leadership of an Elite, 1832–1885* (Chapel Hill, NC, 1986), 34-37.

43. Dale Tomich, *Through the Prism of Slavery*, 83. See also Tomich, "World Slavery and Caribbean Capitalism: The Cuban Sugar Industry, 1760–1868," *Theory and Society* 20:3 (1991), 297-319.

44. See Oscar Zanetti and Alejandro Garcia, *Sugar and Railroads: A Cuban History, 1837–1959* (Chapel Hill, NC, 1987), 19-77.

45. Reinaldo Funes Monzotes, *From Rainforest to Cane Field in Cuba*, 130. See also Maria Portuondo, "Plantation Factories and Technology in Late-Eighteenth-Century Cuba," *Technology and Culture* 44:2 (2003) 244-245; and Manuel Moreno Fraginals, *The Sugar Mill*, 101-107.

46. See Dale Tomich, "World Slavery and Caribbean Capitalism," 297-319.

47. Dale Tomich, *Through the Prism of Slavery*, 85. See also Jonathan Curry-Machado, *Cuban Sugar Industry: Transnational Networks and Engineering Migrants in Nineteenth-Century Cuba* (New York, 2011);

48. See Laird Bergad, *Comparative Histories*, 156-157; and Stanley J. Stein, *Vassouras: A Brazilian Coffee County, 1850–1900* (Cambridge, MA, 1957).

49. Herbert Klein and Francisco Vidal Luna, *Slavery in Brazil* (New York, 2010), 97. See also Rafael de Bivar Marquese and Dale Tomich, "O Vale do Paraiba escravista e a formaçao do Mercado mundial do café no seculo XIX" in Ricardo Salles and Keila Greenberg, eds., *O Brasil Imperio (1808–1889)* (Rio de Janeiro, 2008)

50. See See Leslie Bethell and José Murilo de Carvalho, "1822–1850," 89-94.

51. Laird Bergad, *Comparative Histories*, 160. See also Herbert Klein and Francisco Vidal Luna, *Slavery in Brazil*, 91-100.

52. See Herbert Klein, *Slavery and the Economy of São Paulo, 1750–1850* (Stanford, CA, 2003); Warren Dean, *Rio Claro: A Brazilian Plantation System, 1820–1920* (Stanford, CA, 1976); and Robert Slenes, "The Brazilian Internal Slave Trade, 1850–1888: Regional Economies, Slave Experience, and the Politics of a Peculiar Market" in Walter Johnson, *The Chattel*

Principle: Internal Slave Trades in North America, Brazil, and the West Indies, 1808–1888 (New Haven, CT, 2004), 117-142.

53. See Paul Lovejoy, *Slavery, Commerce, and Production*, 155-205; and A. A. Batran, "The Nineteenth-Century Islamic Revolutions in West Africa" in J. F. Ade Ajayi, ed., *Africa in the Nineteenth Century*, 218-224.

54. Paul Lovejoy, *Slavery, Commerce, and Production*, 175. See also Ralph A. Austen, *Trans-Saharan Africa in World History* (New York, 2010), 66-67.

55. See Mohammed Bashir Salau, "Ribats and the Development of Plantations in the Sokoto Caliphate," *African Economic History* 34 (2006), 23-43.

56. On slavery in Kano, see Sean Stilwell, *Paradoxes of Power: The Kano "Mamluks" and Male Royal Slavery in the Sokoto Caliphate, 1804–1903* (Portsmouth, NH, 2004).

57. See Colleen Kriger, "Textile Production and Gender in the Sokoto Caliphate," *Journal of African History* 34 (1993), 361-401.

58. See Kevin Dawson, "Slave Culture" in Robert L. Paquette and Mark M. Smith, eds., *The Oxford Handbook of Slavery in the Americas* (New York, 2010), 465-488; and James Sidbury, "Resistance to Slavery" in Gad Heuman and Trevor Burnard, eds., *The Routledge History of Slavery* (London, 2011), 204-219.

59. Gwendolyn Midlo Hall, *Slavery and African Ethnicities in the Americas: Restoring the Links* (Chapel Hill, NC, 2007), xv-xvi.

60. See Oliver Pétré-Grenouilleau, "Processes of Exiting the Slave Systems: A Typology" in Enrico Dal Lago and Constantina Katsari, eds., *Slave Systems*, 233-264. See also Matt D. Childs, "Slave Culture" in Gad Heuman and Trevor Burnard, eds., *Routledge History of Slavery*, 170-186.

61. See Ralph A. Austen, *Trans-Saharan Africa in World History*, 92-93.

62. See Robert W. Slenes, "Brazil" in Robert L. Paquette and Mark M. Smith, eds., *Oxford Handbook of Slavery in the Americas*, 123-125.

63. Michael A. Gomez, *Exchanging Our Country Marks: The Transformation of African Identities in the Colonial and Antebellum South* (Chapel Hill, NC, 1998), 194-195.

64. See Albert J. Raboteau, *Slave Religion: The "Invisible Institution" in the Antebellum South* (New York, 1979), 95-150.

65. See Ira Berlin, *The Making of African America: The Four Great Migrations* (New York, 2010), 128-129.

66. See Matt D. Childs and Toyin Fayola, "The Yoruba Diaspora in the Atlantic World: Methodology and Research," in Fayola and Childs, eds., *The Yoruba Diaspora in the Atlantic World* (Bloomington, IN, 2004), 1-16; and Laird Bergad, *Comparative Histories*, 180-187.

67. See Christine Ayorinde, "Santeria in Cuba: Tradition and Transformation," in Toyin Fayola and Matt D. Childs, eds., *The Yoruba Diaspora in the Atlantic World*, 209-230; and Rafael Ocasio, "Dancing to the Beat of Babalu Aye: Santeria and Cuban Popular Culture" in Parick Bellegarde-Smith, ed., *Fragments of Bone: Neo-African Religions in a New World* (Urbana, IL, 2005).

68. See João José Reis, *Death Is a Festival: Funeral Rites and Rebellion in Nineteenth-Century Brazil* (Chapel Hill, NC, 2003), 39-65.

69. See João José Reis, "Candomblé in Nineteenth-Century Bahia: Priests, Followers, Clients," *Slavery & Abolition* 22:1 (2001), 91-115. See also Roger Bastide, *The African Religions of Brazil: Toward a Sociology of the Interpretation of Civilizations* (Baltimore, MD, 1978).

70. See Eugene Genovese, *Roll, Jordan, Roll*; and Genovese, *From Rebellion to Revolution: Afro-American Slave Revolts in the Making of the Modern World* (Baton Rouge, LA, 1979). See also Sylvia R. Frey, "Remembered Pasts: African Atlantic Religions" in Gad Heuman and Trevor Burnard, eds., *Routledge History of Slavery*, 153-169.

71. See Gad Heuman, "Slave Rebellions," in Heuman and Trevor Burnard, eds. *Routledge History of Slavery*, 220-233; and David Geggus, ed., *The Impact of the Haitian Revolution in the Atlantic World* (Columbia, SC, 2001).

72. See John Hope Franklin and Loren Schweninger, *Runaway Slaves: Rebels on the Plantation* (New York, 1999).

73. See Douglas R. Egerton, *Gabriel's Rebellion: the Virginia Slave Conspiracies of 1800 and 1802* (Chapel Hill, NC, 1993).

74. See Daniel Rasmussen, *American Uprising: The Untold Story of America's Largest Slave Revolt* (New York, 2011).

75. Douglas R. Egerton, "Slave Resistance" in Robert L. Paquette and Mark M. Smith, eds., *Oxford Handbook of Slavery in the Americas*, 448. See also Douglas R. Egerton, *He Shall Go Free: The Lives of Denmark Vesey* (New York, 2004).

76. See Kenneth Greenberg, ed., *The Confessions of Nat Turner* (New York, 1995).

77. See Scot French, *The Rebellious Slave: Nat Turner in American Memory* (New York, 2004); and Kenneth Greenberg, ed., *Nat Turner: A Slave Rebellion in History and Memory* (New York, 2003).

78. See Matt D. Childs and Manuel Barcia, "Cuba" in Robert L. Paquette and Mark M. Smith, eds., *Oxford Handbook of Slavery in the Americas*, 96-103.

79. See Laird Bergad, *Comparative Histories*, 201-213. See also Matt D. Childs, *The 1812 Aponte Rebellion in Cuba and the Struggle Against Atlantic Slavery* (Chapel Hill, NC, 2006).

80. See Manuel Barcia, *Seeds of Insurrection: Domination and Resistance in Western Cuban Plantations* (Baton Rouge, LA, 2008); and Robert L. Paquette, *Sugar Is Made with Blood: The Conspiracy of La Escalera and the Conflict between Empires over Slavery in Cuba* (Princeton, NJ, 1988).

81. See Herbert Klein and Francisco Vidal Luna, *Slavery in Brazil*, 209-211.

82. See João José Reis, "Slave Resistance in Brazil: Bahia, 1808–1835," *Luso-Brazilian Review* 25:1 (1988), 111-144; and Stuart B. Schwartz, *Sugar Plantations in the Formation of Brazilian Society: Bahia, 1550–1835* (New York, 1985), 468-488.

83. Laird Bergad, *Comparative Histories*, 231. See also João José Reis, *Slave Rebellions in Brazil: The Muslim Uprising of 1835 in Bahia* (Baltimore, MD, 1993).

84. See Ralph Austen, *Trans-Saharan Africa in World History*, 92-93.

85. See Jennifer Lofkranz, "Protecting Freeborn Muslims," 109-127.

86. Paul Lovejoy, *Slavery, Commerce, and Production*, 257. See also Ian Linden, "Between Two Religions: The Children of Israelites (c.1846–c.1920)" in Elizabeth Isichei, ed., *Varieties of Christian Experience in Nigeria* (London, 1982), 79-98.

87. See Paul Lovejoy, *Slavery, Commerce, and Production*, 55-80.

88. See Paul Lovejoy, *Slavery, Commerce, and Production*, 257-259; João José Reis, *Slave Rebellions in Brazil*; and Patrick Manning, *Slavery and Africa Life: Occidental, Oriental, and African Slave Trades* (Cambridge, MA, 1990), 80-82. See also Sylviane A. Diouf, *Servants of Allah: African Muslims Enslaved in the Americas* (New York, 1998); and Raymundo Nina Rodrigues, *Os Africanos no Brasil* (São Paulo, 1932).

CHAPTER FOUR

1. See Jason W. Moore, "Sugar and the Expansion of the Early Modern World-Economy: Commodity Frontiers, Ecological Transformations, and Industrialization," *Review* 23:3 (2000), 409-433; and Immanuel Wallerstein, *The Capitalist World-Economy* (New York, 1979), 1-36.

2. On the second slavery, see Dale Tomich and Michael Zeuske, "Introduction, The Second Slavery: Mass Slavery, World-Economy, and Comparative Microhistories," *Review* 31:2 (2008), 91-100. On the second serfdom, see Richard Ellie, "Russian Slavery and Serfdom, 1450-1804" in David Eltis and Stanley Engerman, eds., *The Cambridge World History of Slavery*, Vol. 3: *AD 1420–AD 1804* (New York, 2011), 275-296.

3. See the important review of recent comparative historiography in Peter Kolchin, "The South and the World," *Journal of Southern History* 75:3 (2009), 565-580.

4. See Peter Kolchin, *Unfree Labor; American Slavery and Russian Serfdom* (Cambridge, MA, 1987); Shearer Davis Bowman, *Masters and Lords: Mid-Nineteenth-Century U.S. Planters and Prussian Junkers* (New York, 1993); and Enrico Dal Lago, *Agrarian Elites: American Slaveholders and Southern Italian Landowners, 1815–1861* (Baton Rouge, LA, 2005).

5. Michael L. Bush, *Servitude in Modern Times* (London, 1999), ix.

6. Michael L. Bush, *Servitude in Modern Times,* ix.

7. Robert Brenner, "The Rises and Declines of Serfdom in Medieval and Early Modern Europe" in Michael L. Bush, ed., *Serfdom and Slavery: Studies in Legal Bondage* (London, 1996), 273. For the general framework, see Immanuel Wallerstein, "The Rise and Demise of the World Capitalist System: Concepts for Comparative Analysis," *Comparative Studies in Society and History* 16:4 (1974), 387-415.

8. See Michael L. Bush, *Servitude in Modern Times,* 177-185.

9. See Dale Tomich, *Through the Prism of Slavery: Labor, Capital, and World Economy* (Lantham, MA), 2004, 56-74.

10. Dale Tomich and Michael Zeuske, "Introduction," 91.

11. See Eric Wolf, *Europe and the People Without History* (Berkeley, CA, 1982), 267-296; and Robert Gildea, *Barricades and Borders: Europe, 1800–1914* (Oxford, 2003), 148-153.

12. See Carl Levy, "Lords and Peasants" in Stefan Berger, ed., *A Companion to Nineteenth-Century Europe, 1789–1914* (Oxford, 2009), 70-85; and Edgar Melton, "The Russian Peasantries, 1450–1860" in Tom Scott, ed., *The Peasantries of Europe* (Harlow, 1998), 234-257.

13. See Marta Petrusewicz, "Land-Based Modernization and the Culture of Nineteenth-Century Landed Elites" in Enrico Dal Lago and Rick Halpern, eds., *The American South and the Italian Mezzogiorno: Essays in Comparative History* (New York, 2002), 95-111.

14. On southern Italian agriculture, see Enrico Dal Lago, *Agrarian Elites,* 55-77. On Spanish agriculture, see Adrian Schubert, *A Social History of Modern Spain* (London, 1990), 11-15.

15. Edgar T. Thompson, *Plantation Societies, Race Relations, and the Old South: The Regimentation of Populations* (Durham, NC, 1975), 22.

16. See Eric Wolf, *Europe and the People Without History,* 354-383.

17. Marta Petrusewicz, "Land-Based Modernization," 96. See also Lluis Argemì, "Agriculture, Agronomy, and Political Economy: Some Missing Links," *History of Political Economy* 34:2 (2002), 452-460; and Petrusewicz, "Agromania: innovatori agrari nelle periferie europee dell'Ottocento" in Piero Bevilacqua, ed., *Storia dell''agricoltura italiana in età contemporanea,* Vol. 3: *Mercati e Istituzioni* (Venice, 1991), 295-343.

18. See Mark Smith, "Old South Time in Comparative Perspective," *American Historical Review* 101:5 (1996), 1432-1469.

19. See Peter Kolchin, "The South and the World," 565-580.

20. Michael L. Bush, *Servitude in Modern Times,* 3.

21. See Peter Kolchin, *Unfree Labor,* 103-156; and George Fredrickson, "Planters, *Junkers,* and *Pomeschciki,*" *Reviews in American History* 22 (1994), 381-382.

22. See Shearer Davis Bowman, *Masters and Lords,* 42-78; and George Fredrickson, "Planters, *Junkers,* and *Pomeschciki,*" 383-385.

23. See Maria Malatesta, "The Landed Aristocracy during the Nineteenth and Early Twentieth Centuries" in Harmut Kaelble, ed. *The European Way: European Societies during the Nineteenth and Twentieth Centuries* (New York, 2004), 44-67.

24. See Peter Kolchin, "The Process of Confrontation: Patterns of Resistance to Bondage in Nineteenth-Century Russia and the United States," *Journal of Social History* 11:4 (1978), 460-461.

25. August Meitzen's and William Smith's quotes are in Shearer Davis Bowman, *Masters and Lords,* 18.

26. See Jerome Blum, *Lord and Peasant in Russia: From the Ninth to the Nineteenth Century* (Princeton, NJ, 1971), 394-395.

27. Peter Kolchin, *Unfree Labor,* 65. See also Richard L. Rudolph, "Agricultural Structure and Proto-Industrialization in Russia: Economic Development with Unfree Labor," *Journal of Economic History* 45:1 (1985), 47-69.

28. See David Blackbourn, *History of Germany, 1780–1918: The Long Nineteenth Century* (Oxford, 2002), 3-4 and 54-68.

29. Shearer Davis Bowman, *Masters and Lords,* 55.

30. See William W. Hagen, *Ordinary Prussians: Brandeburg Junkers and Villagers, 1500–1840* (New York, 2002), 593-645.

31. On European elites and agronomy, see Marta Petrusewicz, "Agromania"; Lluis Argemì, "Agriculture, Agronomy, and Political Economy"; and Hamish Graham, "Rural Society and Agricultural Revolution" in Stefan Berger, ed., *Nineteenth-Century Europe,* 31-43.

32. On scientific agriculture in Russia, see Olga Elina, "Planting Seeds for the Revolution: The Rise of Russian Agricultural Science, 1860–1920." *Science in Context* 15:2 (2002), 209-237; and Nikolai Riasanovsky, *Russia and the West in the Teaching of the Slavophiles* (Cambridge, MA, 1952).

33. See Shearer Davis Bowman, *Masters and Lords,* 56-63.

34. See Shearer Davis Bowman, "Industrialization and Economic Development in the Nineteenth-Century U.S. South: Some Interregional and Intercontinental Comparative Perspectives" in Susanna Delfino and Michelle Gillespie, eds., *Global Perspectives on Industrial Transformations in the American South* (Columbia, MO, 2005), 76-104.

35. See Carl Levy, "Lords and Peasants," 70-85.

36. See Dale Tomich and Michael Zeuske, "Introduction," 91-100; and Jason Moore, "Sugar and the Expansion of the Early Modern World-Economy," 409-433.

37. See Marta Petrusewicz, *Latifundium: Moral Economy and Material Life in a European Periphery* (Ann Arbor, MI, 1996); and John S. Cohen and Giovanni Federico, *The Growth of Italian Economy, 1820–1960* (New York, 2001), 30-45.

38. See James Simpson, *Spanish Agriculture: The Long Siesta, 1765–1965* (New York, 1995).

39. See Gabriel Tortella, "Agriculture: A Slow-Moving Sector, 1830–1935" in N. Sanchez-Albornoz, ed., *The Economic Modernization of Spain, 1830–1930* (New York, 1987), 42-62.

40. Augusto Placanica, "Il mondo agricolo meridionale: usure, caparre, contratti" in Piero Bevilacqua, ed., *Storia dell"agricoltura italiana in età contemporanea,* Vol. 2: *Uomini e Classi* (Venice, 1990), 203. See also, on sharecropping contracts in nineteenth-century Italy, Enrico Dal Lago, *Agrarian Elites,* 81-82; and Adrian Lyttleton, "Landlords, Peasants, and the Limits of Liberalism" in John A. Davis, ed., *Gramsci and Italy's Passive Revolution* (London, 1979), 126-127.

41. See Salvador Calatayud, Jesus Millan, and M. Cruz Romeo, "Leaseholders in Capitalist Arcadia: Bourgeois Hegemony and Peasant Opportunities in the Valencian Countryside During the Nineteenth Century," *Agricultural History* 17 (2006), 149-166.

42. See Juan Carmona and James Simpson, "The 'Rabassa Morta' in Catalan Viticulture: The Rise and Decline of a Long-Term Sharecropping Contract, 1670s–1920s," *Journal of Economic History* 59:2 (1999), 290-315.

43. Marta Petrusewicz, "Land-Based Modernization," 96-98.

44. See Salvatore Lupo, *Il giardino degli aranci. Il mondo degli agrumi nella storia del Mezzogiorno* (Venice, 1990), 61-69; and Susanna Delfino, "The Idea of Southern Economic Backwardness: A Comparative View of the United States and Italy" in Delfino and Michelle Gillespie, eds., *Global Perspectives,* 106-130.

45. See Carlos Lopez Fernandez and Pedro Marset Campos, "La agricoltura cientifica en la prensa del siglo XIX a traves de los autores autoctonos," *Dynamis* 17 (1997), 239-258; and Eloy Fernandez Clemente, "La enseñanza de la agricoltura en la España del siglo XIX," *Agricoltur y Sociedad* 56 (1990), 113-141.

CHAPTER FIVE

1. Daniel T. Rodgers, "Worlds of Reform" in Gary W. Reichard and Ted Dickson, eds., *America on the World Stage: A Global Approach to U.S. History* (Urbana, IL, 2008), 149; and Robin Blackburn, *The American Crucible: Slavery, Emancipation, and Human Rights* (London, 2011), 348.

2. See, for a model study of this type, Caleb McDaniel, "Our Country Is the World:

Radical American Abolitionists Abroad," Ph.D. Dissertation, Johns Hopkins University, 2006.

3. Two important recent studies that treat the rise of antislavery and abolitionism in transatlantic perspective are Seymour Drescher, *Abolition: A History of Slavery and Antislavery* (New York, 2009); and Robin Blackburn, *The American Crucible,* especially 277-389.

4. See David Brion Davis, *Slavery and Human Progress* (New York, 1985), xvi-xvii; Davis, *Revolutions: Reflections on American Equality and Foreign Liberations* (Cambridge, MA, 1990), 74-75. See also, for the larger context, C. A. Bayly, *The Birth of the Modern World, 1780–1914: Connections and Comparisons* (Oxford, 2004), 125-170; and Thomas Bender, *Nation Among Nations: America's Place in World History* (New York, 2006), 116-181.

5. See Maurice Jackson, "The Rise of Abolition" in Toyin Fayola and Kevin Roberts, eds., *The Atlantic World, 1450–2000* (Bloomington, IN, 2005), 211-248.

6. On some of these points, see Enrico Dal Lago, "Radicalism and Nationalism: Northern 'Liberators' and Southern Laborers in the USA and Italy, 1830–1860" in Enrico Dal Lago and Rick Halpern, eds., *The American South and the Italian Mezzogiorno: Essays in Comparative History* (New York, 2002), 197-214; and Roland Sarti, "La democrazia radicale: uno sguardo reciproco tra Stati Uniti e Italia," in Maurizio Ridolfi, ed., *La democrazia radicale nell'Ottocento europeo. Forme della politica, modelli culturali, riforme sociali* (Milan, 2005), 133-158.

7. Carl Guarneri, *America in the World: United States History in Global Context* (Boston, 2007), 153.

8. Christopher Schmidt-Nowara, "Empires Against Emancipation: Spain, Brazil, and the Abolition of Slavery," *Review* 31: 2 (2008), 103.

9. See Richard Newman, *The Transformation of Abolitionism: Fighting Slavery in the Early Republic* (New York, 2002), 1-15; Richard J. M. Blackett, *Building an Antislavery Wall: Black Americans in the Atlantic Antislavery Movement* (Baton Rouge, LA, 1983); and Timothy P. McCarthy and John Stauffer, eds., *Prophets of Protest: Reconsidering the History of American Abolitionism* (Boston, 2006).

10. William Lloyd Garrison, "Editorial," *The Liberator,* January 1, 1831. See also Henry Mayer, *All on Fire: William Lloyd Garrison and the Abolition of Slavery* (New York, 1998), 97-126.

11. See William Lloyd Garrison, "The American Anti-Slavery Society's Declaration of Sentiments (1833)" in Rick Halpern and Enrico Dal Lago, eds., *Slavery and Emancipation* (Oxford, 2002), 299-301, See also, on American abolitionism in the 1830s, Herbert Aptheker, *Abolitionism: A Revolutionary Movement* (New York, 1989); Ronald Walters, *The Anti-Slavery Appeal: American Abolitionism after 1830* (New York, 1976); and Merton Dillon, *The Abolitionists: The Growth of a Dissenting Minority* (New York, 1976).

12. Thomas Bender, *Nation Among Nations,* 153. See also James B. Stewart, *Holy Warriors: The Abolitionists and American Slavery* (New York, 1996), 51-74.

13. Ian Tyrrell, *Transnational Nation: United States History in Global Perspective since 1789* (New York, 2007), 49. See also Richard J. M. Blackett, "And There Shall Be No More Sea: William Lloyd Garrison and the Transatlantic Abolitionist Movement" in James B. Stewart, ed., *William Lloyd Garrison at Two Hundred* (New Haven, CT, 2008), 113-141.

14. See Caleb McDaniel, "Our Country Is the World"; Henry Mayer, *All On Fire,* 150-151.

15. See David Brion Davis, "Declaring Equality: Sisterhood and Slavery" in Kathryn Kish Sklar and James B. Stewart, eds., *Women's Rights and Transatlantic Slavery in the Era of Emancipation* (New Haven, CT, 2007), 3-18.

16. See Aileen S. Kraditor, *Means and Ends in American Abolitionism: Garrison and His Critics on Strategy and Tactics, 1834–1850* (New York, 1969).

17. See Betty Fladeland, *Men and Brothers: Anglo-American Antislavery Cooperation* (Urbana, IL, 1972); David Brion Davis, *The Problem of Slavery in the Age of Revolution, 1770–1823* (Ithaca, NY, 1975); and Edward B. Rugemer, *The Problem of Emancipation: The Caribbean Roots of the American Civil War* (Baton Rouge, LA, 2008).

18. See Seymour Drescher, *Econocide: British Slavery in the Era of Abolition,* 2nd ed. (Chapel Hill, NC, 2010); David Beck Ryden, *West Indian Slavery and British Abolition, 1783–1807* (New York, 2009); and Adam Hochshild, *Bury the Chains: The British Struggle to Abolish Slavery* (New York, 2005), 323-327.

19. See Kenneth Morgan, *Slavery and the British Empire: From Africa to America* (New York, 2007), 184-198; and William Green, *British Slave Emancipation: The Sugar Colonies and the Great Experiment, 1830–1865* (New York, 1976).

20. Howard R. Temperley, "Abolition and Antislavery Movements: Great Britain" in Paul Finkelman and Joseph Miller, eds. *MacMillan Encyclopedia of World Slavery* (New York, 1997), 10.

21. See Nicholas Draper, *The Price of Emancipation: Slave-Ownership, Compensation and British Society at the End of Slavery* (New York, 2010); and Howard R. Temperley, *British Antislavery, 1833–1870* (London, 1972).

22. David Brion Davis, *Inhuman Bondage: The Rise and Fall of Slavery in the New World* (New York, 2006), 238; Seymour Drescher, *The Mighty Experiment: Free Labor versus Slavery in British Emancipation* (New York, 2002); Deborah Bingham Van Broekhoven, "Perspectives on Slavery: Opposition to Slavery" in Paul Finkelman and Joseph Miller, eds., *Encyclopedia of World Slavery,* 691. For the Atlantic context, see Jeffrey Kerr-Ritchie, *Rites of August First: Emancipation Day in the Black Atlantic World* (Baton Rouge, LA, 2007).

23. Robin Blackburn, *American Crucible,* 348. See also Seymour Drescher, *Abolition,* 267-285; Simon Morgan, "The Anti-Corn Law League and British Anti-Slavery in Transatlantic Perspective, 1838–1846," *Historical Journal* 52:1 (2009), 87-107; and Marika Sherwood, "Britain, the Slave Trade, and Slavery, 1808–1843," *Race & Class* 46:2 (2004): 54-77.

24. Maurice J. Bric, "Daniel O'Connell and the Debit on Anti-Slavery, 1820–1850" in Tom Dunne and Laurence M. Geary, eds., *History and the Public Sphere: Essays in Honor of John A. Murphy* (Cork, 2005), 72.

25. See John F. Quinn, "Expecting the Impossible? Abolitionist Appeals to the Irish in Antebellum America," *New England Quarterly* 82:4 (2009), 667-710.

26. See Douglas Riach, "Richard Davis Webb and Antislavery in Ireland" in Lewis Perry and Michael Fellman, eds., *Antislavery Reconsidered: New Perspectives on the Abolitionists* (Baton Rouge, LA, 1979), 149-167.

27. Nini Rodgers, *Ireland, Slavery, and Anti-Slavery, 1612–1865* (New York, 2005), 259. See also Patrick M. Geoghegan, *King Dan: The Rise of Daniel O'Connell, 1775–1829* (Dublin, 2008).

28. See Patrick M. Geoghegan, *Liberator: The Life and Death of Daniel O'Connell, 1830–1847* (Dublin, 2010), 197-211.

29. On these issues, see Douglas C. Riach, "Ireland and the Campaign Against American Slavery, 1830–1860," Ph.D. Dissertation, University of Edinburgh, 1975. See also Douglas C. Riach, "Daniel O'Connell and American Anti-Slavery," *Irish Historical Studies* 20 (1976), 3-25.

30. See Robin Blackburn, *The Overthrow of Colonial Slavery, 1776–1848* (London, 1988), 161-264.

31. For the context of these issues, see C. A. Bayly, *Birth of the Modern World.*

32. See also M. P. Kielstra, *The Politics of Slave Suppression in Britain and France, 1814–1848* (London, 2000).

33. See Lawrence Jennings, *French Reactions to British Emancipation* (New York, 1988); and David Todd, "A French Imperial Meridian, 1814–1870," *Past & Present* 210 (2011), 154-186.

34. See Olivier Pétré-Grenouilleau, "Abolitionnisme et democratization" in Pétré-Grenouilleau, ed., *Abolir l'esclavage: Un reformisme a l'epreuve (France, Portugal, Suisse, XVIIIe-XIXe sicles)* (Paris, 2008), 7-23.

35. See Lawrence Jennings, *French Anti-Slavery: The Movement for the Abolition of Slavery in France, 1802–1848* (New York, 2000).

36. See Nelly Schmidt, *Victor Schoelcher et l'abolition de l'esclavage* (Paris, 1994).

37. Thomas Benjamin, *The Atlantic World: Europeans, Africans, Indians and Their Shared History, 1400–1900* (New York, 2009), 641.

38. William Lloyd Garrison, "To the Public," *The Liberator,* January 1, 1831. See also John L. Thomas, *The Liberator. William Lloyd Garrison: A Biography* (Boston, MA, 1963); and James B. Stewart, *Holy Warriors,* 52-56.

39. See Richard J. M. Blackett, "And There Shall Be No More Sea," 13-40.

40. See William E. Cain, "Introduction: William Lloyd Garrison and the Fight Against Slavery" in Cain, ed., *William Lloyd Garrison and the Fight Against Slavery: Selections from The Liberator* (New York, 1995), 4-33.

41. William Lloyd Garrison, "The American Anti-Slavery Society's Declaration of Sentiments (1833)" in Rick Halpern and Enrico Dal Lago, eds., *Slavery and Emancipation,* 300. See also James B. Stewart, *William Lloyd Garrison and the Challenge of Emancipation* (Arlington Heights, IL, 1992).

42. On these issues, see Caleb McDaniel, "Our Country Is the World," 271-329.

43. On Garrison and women's activism, see Lois A. Brown, "William Lloyd Garrison and Emancipatory Feminism in Nineteenth-Century America" in James B. Stewart, ed., *William Lloyd Garrison at Two Hundred,* 41-76.

44. See Henry Mayer, *All On Fire.*

45. See George Holyoake, *Sixty Years of an Agitator's Life* (London, 1892).

46. Emily Venturi, "W. H. Ashurst: Brief Record of His Life" in *Mr. Morris 80th Birthday Celebration, with a Brief Record of the Life of William Henry Ashurst* (London, 1904); and Matthew Lee, "Ashurst, William Henry," *Oxford Dictionary of National Biography* (Oxford, 2004), available at www.oxforddnb.com.

47. See Maura O'Connor, *The Romance of Italy and the English Political Imagination* (Houndmills, 1998), 71-76.

48. See Enrico Verdecchia, *Londra dei cospiratori.* L'esilio londinese dei padri del Risorgimento (Milan, 2010), 270-271.

49. See Caleb McDaniel, "Our Country Is the World."

50. Gearoid O'Tuathaigh, "O'Connell, Daniel" in James McGuire and James Quinn, eds., *Dictionary of Irish Biography Online* (Cambridge and Dublin, 2009), available at http://dib.cambridge.org. See also Oliver McDonagh, *O'Connell: The Life of Daniel O'Connell, 1775–1847* (Dublin, 1991).

51. See Fergus O'Ferrall, *Daniel O'Connell* (Dublin, 1981); and Patrick Geoghegan, *King Dan,* 187-208.

52. See Victor Conzemius, "The Place of Daniel O'Connell in the Liberal Catholic Movement of the Nineteenth Century" in Donal McCartney, ed., *The World of Daniel O'Connell* (Dublin, 1980).

53. See Patrick Geoghegan, *King Dan,* 229-270; and Angela Murphy, *American Slavery and Irish Freedom: Abolition, Immigrant Citizenship, and the Transatlantic Movement for Irish Repeal* (Baton Rouge, LA, 2010).

54. See Patrick Geoghegan, *The Liberator,* 197-211.

55. See especially Christine Kinealy, *Daniel O'Connell and the Anti-Slavery Movement: The Saddest People the Sun Sees* (London, 2010).

56. See Nini Rodgers, *Ireland, Slavery, and Anti-Slavery,* 259-277; Maurice Bric, "Daniel O'Connell," 69-82; and Douglas Riach, "O'Connell and Slavery" in Donal McCartney, ed. *The World of Daniel O'Connell,* 175-185.

57. Giuseppe Mazzini, "Istruzione generale della Giovine Italia" in Franco Della Peruta, ed., *Scrittori politici dell'Ottocento,* Vol. I: *Giuseppe Mazzini e i democratici* (Milan, 1966), 325. See also Roland Sarti, "Giuseppe Mazzini and his Opponents" in John A. Davis, ed., *Italy in the Nineteenth Century, 1796–1800* (Oxford, 2000), 74-107; and Denis Mack Smith, *Mazzini* (New Haven, CT, 1994).

58. See Roland Sarti, "Giuseppe Mazzini and Young Europe" in C. A. Bayly and Eugenio Biagini, eds., *Giuseppe Mazzini and the Globalisation of Democratic Nationalism, 1830–1920*

(Oxford, 2008), 275-298; and Roland Sarti, *Mazzini: A Life for the Religion of Politics* (Westport, CT, 1997), 71-94.

59. See Gregory Claeys, "Mazzini, Kossuth, and British Radicalism, 1848–1854," *Journal of British Studies* 28:3 (1989), 225-261; William Lloyd Garrison, "Introduction" in Garrison, ed., *Joseph Mazzini: His Life, Writings, and Political Principles* (New York, 1972); and Joseph Rossi, *The Image of America in Mazzini's Writings* (Madison, WI, 1954).

60. Douglas H. Maynard, "The World's Anti-Slavery Convention of 1840," *Mississippi Valley Historical Review* 47 (1960), 452.

61. See Catherine Hall, *Civilising Subjects: Metropole and Colony in the English Imagination, 1830–1867* (Oxford, 2002), 329-334.

62. See Caleb McDaniel, "Our Country Is the World," 29-56.

63. See Donald R. Kennon, "An Apple of Discord: The Woman Question at the World's Anti-Slavery Convention," *Slavery & Abolition* 5 (1984), 244-266.

64. British and Foreign Anti-Slavery Society, *Proceedings of the General Anti-Slavery Convention ... Held in London ... 1840* (London, 1840), 13.

65. British and Foreign Anti-Slavery Society, *Proceedings of the General Anti-Slavery Convention,* 301.

66. See James B. Stewart, *Holy Warriors,* 151-180; Bruce Levine, *Half Slave & Half Free: The Roots of the Civil War* (New York, 2002, 2nd ed.); and Orville Vernon Burton, *The Age of Lincoln* (New York, 2008).

67. C. A. Bayly, *Birth of the Modern World,* 158.

68. On these issues, see Caleb McDaniel, "Our Country Is the World," 271-329. See also Timothy M. Roberts, *Distant Revolutions: 1848 and the Challenge to American Exceptionalism* (Charlottesville, VA, 2009).

69. Roland Sarti, "Giuseppe Mazzini and his Opponents."

70. See Lucy Riall, *Risorgimento: The History of Italy from Napoleon to Nation-State* (New York, 2009).

71. On *histoire croisée* (entangled history), see Michael Werner and Benedicte Zimmermann, "Beyond Comparison: Histoire Croisée and the Challenge of Reflexivity," *History and Theory* 45 (2006), 30-50.

CHAPTER SIX

1. See Thomas Bender, *Nation Among Nations: America's Place in the World* (New York, 2006); Carl Guarneri, *America in the World: United States History in Global Context* (New York, 2007); and Ian Tyrrell, *Transnational Nation: United States History in Global Perspective since 1789* (New York, 2007).

2. Michael Geyer and Charles Bright, "Global Violence and Nationalizing Wars in Eurasia and America: The Geopolitics of War in the Mid-Nineteenth Century," *Comparative Study in Society and History* 38 (1996), 620; and Carl Guarneri, *America in the World,* 158.

3. See Eric Foner, *Nothing but Freedom: Emancipation and Its Legacy* (Baton Rouge, LA, 1983); Steven Hahn, "Class and State in Postemancipation Societies: Southern Planters in Comparative Perspective," *American Historical Review* 95:1 (1990), 75-98; Peter Kolchin, *A Sphinx on the American Land: The Nineteenth-Century South in Comparative Perspective* (Baton Rouge, LA, 2003); Rebecca Scott, *Degrees of Freedom: Louisiana and Cuba after Slavery* (Baton Rouge, LA, 2005); and Stanley L. Engerman, "Emancipation Schemes: Different Ways of Ending Slavery" in Enrico Dal Lago and Constantina Katsari, eds., *Slave Systems: Ancient and Modern* (New York, 2008), 265-282.

4. See Brian Shoen, *The Fragile fabric of Union: Cotton, federal Politics, and the Global Origins of the Civil War* (Baltimore, MD, 2009); and Christopher Schmidt-Nowara, *Slavery, Freedom, and Abolition in Latin America and the Atlantic World* (Albuquerque, NM, 2011).

5. Dale Tomich and Michael Zeuske, "Introduction, The Second Slavery: Mass Slavery, World-Economy, and Comparative Microhistories," *Review* 31:2 (2008), 91.

6. See Susan-Mary Grant, *The War for a Nation: The American Civil War* (London, 2010).

7. See Robert Bonner, *Mastering America: Southern Slaveholders and the Crisis of American Nationhood* (New York, 2009); and Stephanie McCurry, *Confederate Reckoning: Power and Politics in the Civil War South* (Cambridge, MA, 2010).

8. Carl Degler, "Thesis, Antithesis, Synthesis: The South, the North, and the Nation," *Journal of Southern History* 53:1 (1987), 18. Modern critics have disputed Hegel's actual paternity of Hegelian dialectic; see Walter Kaufman, *Hegel: A Reinterpretation* (New York, 1966).

9. Eric Foner, *Give Me Liberty! An American History,* Vol. I (New York, 2011), 557.

10. See David Potter, "Civil War" in C. Van Woodward, ed., *The Comparative Approach to American History* (New York, 1968), 135-145.

11. See Richard H. Sewell, *Ballots for Freedom: Antislavery Politics in the United States, 1848–1865* (New York, 1976).

12. See Don Fehrenbacher, *The Slaveholding Republic: An Account of the United States Government's Relation to Slavery* (New York, 2001).

13. See John Ashworth, "Free Labor, Slave Labor, and the Slave Power: Republicanism and the Republican Party in the 1850s" in Melvyn Stokes and Stephen Conway, eds., *The Market Revolution in America: Social, Political, and Religious Expressions, 1800–1880* (Charlottesville, VA, 1997), 138-141.

14. See Eric Foner, *Free Soil, Free Labor, Free Men: The Ideology of the Republican Party before the Civil War* (New York, 1970).

15. See Eric Foner, *Free Soil, Free Labor, Free Men.*

16. See Peter J. Parish, *The North and the Nation in the Era of the Civil War* (New York, 2003). See also William Gienapp, *The Origins of the Republican Party, 1852–1856* (New York, 1987).

17. Peter J. Parish, *The North and the Nation,* 118.

18. See Sean Wilentz, *The Rise of American Democracy: Jefferson to Lincoln* (New York, 2005), 701-702. On the relationship between American republicanism and liberalism, see David Ericson, *The Shaping of American Liberalism: The Debates over Ratification, Nullification, and Slavery* (Chicago, 1993), 2-3.

19. Abraham Lincoln, "The "House Divided" speech (1858)" in Michael P. Johnson, ed., *Abraham Lincoln, Slavery, and the Civil War: Selected Writings and Speeches* (New York, 2001), 125; and Ian Tyrrell, *Transnational Nation,* 87. See also Eric Foner, *A Fiery Trial: Abraham Lincoln and American Slavery* (New York, 2010), 99-103; and Richard Carwardine, *Lincoln: A Life of Purpose and Power* (New York, 2007), 76-78.

20. Mark Hewitson, "Conclusion," and Stefan Berger, "Ethnic Nationalism per Excellence?" both in Timothy Bancroft and Mark Hewitson, eds., *What Is a Nation? Europe, 1789–1914* (Oxford, 2006), 354, 43. See also, on civic and ethnic nationalism, Liah Greenfeld, *Nationalism: Five Roads to Modernity* (Cambridge, MA, 1995).

21. On liberalism and nationalism, see Mark Haugaard, "Nationalism and Liberalism" in Gerard Delanty and Krishan Kumar, eds., *The SAGE Handbook of Nations and Nationalism* (London, 2006), 345-356.

22. Stefan Berger, "National Movements" in Berger, ed., *A Companion to Nineteenth-Century Europe, 1789–1914* (Oxford, 2009), 186. See also Patrick Lawrence, *Nationalism in Europe, 1780–1850* (London, 2004); and Eric J. Hobsbawm, *Nations and Nationalism since 1780: Programme, Myth, Reality* (Cambridge, 1994).

23. See Roberto Balzani, "Cavour e le vie della guerra" in Mario Isnenghi and Eva Cecchinato, eds., *Fare l'Italia: Unità e disunità nel Risorgimento* (Turin, 2008), 342-356; and Alberto Mario Banti, *Il Risorgimento italiano* (Rome-Bari, 2007).

24. See Rosario Romeo, *Vita di Cavour* (Rome-Bari, 1985).

25. See David Brading, *The First America: The Spanish Monarchy, Creole Patriots, and the Liberal State, 1492–1867* (New York, 1991); and Jaime E. Rodriguez, *The Independence of Spanish America* (New York, 1998).

26. Will Fowler, *Latin America since 1780* (London, 2008), 38. See also John Lynch, ed., *Latin American Revolutions, 1808–1826: Old and New World Regimes* (Norman, OK, 1994).

27. See David Bushnell and Neil Macaulay, *The Emergence of Latin America in the Nineteenth Century* (New York, 1983); and John Lynch, *Caudillos in Spanish America, 1800–1850* (Oxford, 1992).

28. See David Bushnell and Neil Macaulay, *The Emergence of Latin America in the Nineteenth Century*; and Rebecca Earle, ed., *Rumors of Wars: Civil Conflict in Nineteenth-Century Latin America* (London, 2000).

29. For some important comparative points, see Nicholas and Peter Onuf, *Nations, Markets, and Wars: Modern History and the American Civil War* (Charlottesville, VA, 2006); Don H. Doyle and Marco Pamplona, eds., *Nationalism in the New World* (Athens, GA, 2006); and James Dunkerley, *Americana: The Americas in the World Around 1850* (London, 2000).

30. See Seymour Drescher, *Abolition: A History of Slavery and Antislavery* (New York, 2009), 267-333.

31. Seymour Drescher, *Abolition,* 273-274, 279.

32. Seymour Drescher, *Abolition,* 279. See also Edward Rugemer, *The Problem of Emancipation: The Caribbean Roots of the American Civil War* (Baton Rouge, LA, 2008), 258-290; and Gerald Horne, *The Deepest South: The United States, Brazil, and the African Slave Trade* (New York, 2007).

33. See S. Daget, "The Abolition of the Slave Trade" in J. F. Ade Ajayi, ed., *UNESCO General History of Africa,* Vol. VI: *Africa in the Nineteenth Century until the 1880s* (Berkeley, CA, 1998), 27-38.

34. See S. Daget, "The Abolition of the Slave Trade," 27-38.

35. See Sue Peabody and Keila Grinberg, "Introduction: Slavery, Freedom, and the Law" in Peabody and Grinberg, eds., *Slavery, Freedom, and the Law in the Atlantic World* (New York, 2007), 1-28.

36. See Michael L. Bush, *Servitude in Modern Times* (Cambridge, 2000), 177-199.

37. See Ira Berlin et al., *Slaves No More: Three Essays on Emancipation and the Civil War* (New York, 1992), 1-76.

38. See Paul Finkelman, "Lincoln and the Preconditions for Emancipation: The Moral Grandeur of a Bill of Lading" in William A. Blair and Karen Fischer Younger, eds., *Lincoln's Proclamation: Emancipation Reconsidered* (Chapel Hill, NC, 2009), 13-44; and Eric Foner, *A Fiery Trial,* 166-206.

39. See Michael Vorenberg, *Final Freedom: The Civil War, the Abolition of Slavery and the Thirteenth Amendment* (New York, 2001), 23-35.

40. Abraham Lincoln, "The Emancipation Proclamation (1863)" in Rick Halpern and Enrico Dal Lago, eds., *Slavery and Emancipation* (Oxford, 2002), 380. See also Michael Vorenberg, "Abraham Lincoln's 'Fellow Citizens'—Before and After Emancipation" in William A. Blair and Karen Fischer Younger, eds., *Lincoln's Proclamation,* 151-169.

41. See Steven Hahn, *The Political Worlds of Slavery and Freedom* (Cambridge, MA, 2009), 55-114; and Stephanie McCurry, *Confederate Reckoning,* 218-262.

42. See Herman Belz, "The Constitution, the Amendment Process, and the Abolition of Slavery" in Harold Holzer and Sarah Vaughn Gabbard, eds., *Lincoln and Freedom: Slavery, Emancipation, and the Thirteenth Amendment* (Carbondale, IL, 2007), 160-179.

43. See Christopher Schmidt-Nowara, *Empire and Antislavery: Spain, Cuba, and Puerto Rico, 1833–1874* (Pittsburgh, PA, 1999); and Matthew Pratt Guterl, *American Mediterranean: Southern Slaveholders in the Age of Emancipation* (Cambridge, MA, 2008).

44. See Matt D. Childs and Manuel Barcia, "Cuba" in Mark M. Smith and Robert L. Paquette, eds., *The Oxford Handbook of Slavery in the Americas* (New York, 2010), 103-106; and Ada Ferrer, *Insurgent Cuba: Race, Nation, and Revolution, 1868–1898* (Chapel Hill, NC, 1999).

45. Matt D. Childs and Manuel Barcia, "Cuba," 103. See also Ada Ferrer, "Armed Slaves and Anticolonial Insurgency in Late Nineteenth-Century Cuba" in Christopher L. Brown and

Philip D. Morgan, eds., *Arming Slaves: From Classical Times to the Modern Age* (New Haven, CT, 2006), 304-329.

46. See Rebecca J. Scott, *Slave Emancipation in Cuba: The Transition to Free Labor, 1860–1899* (Princeton, NJ, 1985).

47. See Leslie Bethell, *The Abolition of the Brazilian Slave Trade: Britain, Brazil, and the Slave Trade Question, 1807–1864* (New York, 1970).

48. Robert W. Slenes, "Brazil" in Robert Paquette and Mark Smith, eds., *Slavery in the Americas,* 124. See also Jeffrey Needell, *The Party of Order: The Conservatives, the State, and Slavery in the Brazilian Monarchy, 1831–1871* (Stanford, CA, 2006).

49. See Christopher Schmidt-Nowara, "Empires against Emancipation: Spain, Brazil, and the Abolition of Slavery," *Review* 31:2 (2008), 114-115.

50. See Emilia Viotti da Costa, "1870–1889" in Leslie Bethell, ed., *Brazil: Empire and Republic, 1822–1930* (New York, 1989), 161-213.

51. See Herbert S. Klein and Francisco Vidal Luna, *Slavery in Brazil* (New York, 2010), 295-320.

52. See Robert Conrad, *The Destruction of Brazilian Slavery, 1850–1888* (Berkeley, CA, 1972).

53. See Carl Levy, "Lords and Peasants" in Stefan Berger, ed., *A Companion to Nineteenth-Century Europe,* 70-85.

54. See Jerome Blum, *The End of the Old Order in Europe* (Princeton, NJ, 1978).

55. Michael L. Bush, *Servitude in Modern Times,* 185.

56. See David Field, *The End of Serfdom: Nobility and Bureaucracy in Russia, 1855–1861* (Cambridge, MA, 1976).

57. Vera Tolz, "Russia: Empire or Nation-State in the Making?" in Bancroft and Hewitson, eds., *What Is a Nation?,* 307. See also Boris N. Mironov, "When and Why Was the Russian Peasantry Emancipated?" in Michael L. Bush, ed., *Serfdom & Slavery: Studies in Legal Bondage* (London, 1996), 323-347.

58. See Peter Kolchin, "Some Controversial Questions Concerning Nineteenth-Century Slavery and Serfdom" in Michael L. Bush, ed., *Serfdom and Slavery,* 42-68.

59. Peter Kolchin, "After Serfdom: Russian Emancipation in Comparative Perspective" in Stanley L. Engerman, ed., *Terms of Labor: Slavery, Serfdom, and Free Labor* (Stanford, CA, 1999), 92, and, in general, 88-95. See also David Moon, *The Abolition of Serfdom in Russia, 1762–1902* (Harlow, 2001).

60. Michael L. Bush, *Servitude in Modern Times,* 177.

61. See Stanley L. Engerman, "Emancipation Schemes," 265-282.

62. Carl Degler, "One Among Many: The United States and National Unification" in Gabor Boritt, ed., *Lincoln: The War President* (New York, 1992), 106. See also David Potter, "Civil War," 135-145.

63. Eric J. Hobsbawm, "Introduction: Inventing Traditions" in Hobsbawm and Terence Ranger, eds., *The Invention of Tradition* (Cambridge, 1983), 13. See also Hobsbawm, *Nations and Nationalism since 1780: Programme, Myth, and Reality* (Cambridge, 1990); Ernest Gellner, *Nations and Nationalism* (London, 1983); Benedict Anderson, *Imagined Communities: Reflections on the Origins and Spread of Nationalism* (London, 1983); James McPherson, *Is Blood Thicker than Water? Crises of Nationalism in the Modern World* (New York, 1999); Drew Faust, *The Creation of Confederate Nationalism: Ideology and Identity in the Civil War South* (Baton Rouge, LA, 1988); and Susan-Mary Grant, *North over South: Northern Nationalism and America Identity in the Antebellum Era* (Lawrence, KS, 2000).

64. David Potter, "Civil War," 138, 143.

65. See Timothy M. Roberts, *Distant Revolutions: 1848 and the Challenge to American Exceptionalism* (Charlottesville, VA, 2009).

66. In general, on Lincoln during the Civil War, see Richard Carwardine, *Abraham Lincoln,* 191-310; and Philip S. Paludan, *The Presidency of Abraham Lincoln* (Lawrence, KS, 1994).

67. See Howard Belz, *Abraham Lincoln, Constitutionalism, and Equal Rights in the Civil War Era* (New York, 1998).

68. Melinda Lawson, *Patriot Fires: Forging a New American Nationalism in the Civil War North* (Lawrence, KS, 2002), 161; Lincoln's quotation is in Richard P. Basler, ed., *The Collected Works of Abraham Lincoln,* Vol. 7, 395. See also Richard Carwardine, "Abraham Lincoln, the Presidency, and the Mobilization of Union Sentiment" and Susan-Mary Grant, "From Union to Nation? The Civil War and the Development of American Nationalism," both in Grant and Brian Holden Reid, eds., *The American Civil War: Explorations and Reconsiderations* (London, 2008), 68-97, 333-357.

69. Abraham Lincoln, "Gettysburg Address" in Michael P. Johnson, ed., *Abraham Lincoln, Slavery, and the Civil War,* 263. See also Gary Wills, *Lincoln at Gettysburg: The Words that Remade America* (New York, 1992).

70. See Harry V. Jaffa, *A New Birth of Freedom: Abraham Lincoln and the Coming of the Civil War* (New York, 2000), 73-152; and David Herbert Donald, *Lincoln* (New York, 1995), 460-466.

71. See Luciano Cafagna, *Cavour* (Bologna, 1999); and Adriano Viarengo, *Cavour* (Salerno, 2010).

72. Camillo Cavour's quotation is in Denis Mack Smith, ed., *The Making of Italy* (London, 1986), 326-327. See also Lucy Riall, *Risorgimento: The History of Italy from Napoleon to Nation State* (New York, 2009); and Rosario Romeo, *Dal Piemonte sabaudo all'Italia liberale* (Rome-Bari, 1963).

73. For comparisons and connections, see Tiziano Bonazzi, "Postfazione: la Guerra Civile americana e la "nazione universale" in Tiziano Bonazzi e Carlo Galli, eds., *La Guerra Civile Americana vista dall'Europa* (Bologna, 2004), 463-502; Raimondo Luraghi, *Storia della Guerra Civile Americana* (Turin, 1966); Eugenio Biagini, "The Principle of Humanity: Lincoln in Germany and Italy, 1859–1865," in Richard Carwardine and Jay Sexton, eds., *The Global Lincoln* (New York, 2011), 76-94; and Don H. Doyle, *Nations Divided: America, Italy, and the Southern Question* (Athens, GA, 2003).

74. Carl Guarneri, *America in the World,* 158. See also Eugenio Biagini, "The Principle of Humanity," 78-79; and Paul Quigley, "Secessionists in an Age of Secession: The Slave South in Transatlantic Perspective" in Don H. Doyle, ed., *Secession as an International Phenomenon: From America's Civil War to Contemporary Separatist Movements* (Athens, GA, 2010), 158-159.

75. Peter J. Parish, *The North and the Nation,* 118.

76. See Eric J. Hobsbawm, *The Age of Capital, 1848–1875* (London, 1975); and C.A. Bayly, *The Birth of the Modern World, 1780–1914: Connections and Comparisons* (Oxford, 2004).

CONCLUSION

1. See Frank Tannenbaum, *Slave and Citizen* (New York, 1946); and Robin Blackburn, *The American Crucible: Slavery, Emancipation, and Human Rights* (London, 2011).

2. See Anthony Kaye, "The Second Slavery: Modernity in the Nineteenth-Century South and the Atlantic World," *Journal of Southern History* 75:3 (2009), 627-650.

3. See Laird Bergad, *The Comparative Histories of Brazil, Cuba, and the United States* (New York, 2007).

4. See Peter Kolchin, *A Sphinx on the American Land: The Nineteenth-Century South in Comparative Perspective* (Baton Rouge, LA, 2003); and Michael L. Bush, *Servitude in Modern Times* (London, 2000).

5. See David Brion Davis, *Inhuman Bondage: The Rise and Fall of Slavery in the New World* (New York, 2006); Seymour Drescher, *Abolition: A History of Slavery and Antislavery* (New York, 2009); and Robin Blackburn, *American Crucible.*

6. See C. A. Bayly, *The Birth of the Modern World, 1780–1914: Connections and Comparisons* (Oxford, 2004); Thomas Bender, *A Nation Among Nations: America's Place in World*

History (New York, 2007); and Nicholas and Peter Onuf, *Nations, Markets, and War: Modern History and the American Civil War* (Charlottesville, VA, 2006).

7. Robin Blackburn, *American Crucible,* 388.

8. Eric Foner, *Forever Free: The Story of Emancipation and Reconstruction* (New York, 2005), 202.

9. Robin Blackburn, *American Crucible,* 462-463.

10. Edward Rugemer, *The Problem of Emancipation: The Caribbean Roots of the Civil War* (Baton Rouge, LA, 2008), 301.

11. See Paul Lovejoy, *Slavery, Commerce, and Production in the Sokoto Caliphate of West Africa* (Trenton, NJ, 2005), 43-48.

12. See Peter Kolchin, "After Serfdom: Russian Emancipation in Comparative Perspective" in Stanley L. Engerman, ed., *Terms of Labor: Slavery, Serfdom, and Free Labor* (Stanford, CA, 1999), 87-115.

13. Carl Levy, "Lords and Peasants" in Stefan Berger, ed., *A Companion to Nineteenth-Century Europe* (Oxford, 2009), 76. See also, for comparisons, Steven Hahn, "Class and State in Post Emancipation Societies: Southern Planters in Comparative Perspective," *American Historical Review* 95:1 (1990), 75-98.

BIBLIOGRAPHICAL ESSAY

The purpose of this essay is to provide a short list that combines the most important works presented in the endnotes of each chapter with other bibliographical suggestions, clearly divided into groups according to their common themes. The essay is intended as a guide to the very large and complex literature on American slavery in comparative and transnational perspective. Its presentation and organization of the scholarly works reflect the double focus of the book, with its Atlantic and Euro-American approach to the U.S. peculiar institution.

GENERAL STUDIES AND HISTORIOGRAPHY

General studies that have placed slavery in the American South within a general world history framework include David Brion Davis, *The Problem of Slavery in Western Culture* (New York, 1966), *The Problem of Slavery in the Age of Revolutions, 1770–1825* (Ithaca, NY, 1975), and *Slavery and Human Progress* (New York, 1984); Immanuel Wallerstein, *The Modern World-System,* 4 vols. (New York, 1974–2011); Orlando Patterson, *Slavery and Social Death: A Comparative Study* (Cambridge, MA, 1982); Eric Wolf, *Europe and the People Without History* (Berkeley, CA, 1982); Philip Curtin, *The Rise and Fall of the Plantation Complex: Essays in Atlantic History* (New York, 1990); Dale Tomich, *Through the Prism of Slavery: Labor, Capital, and World Economy* (Lantham, MD, 2004); C. A. Bayly, *The Birth of the Modern World, 1780–1914: Global Connections and Comparisons* (Oxford, 2004); James Belich, *Replenish the Earth: The Settler Revolution and the Rise of the Anglo-World, 1783–1939* (New York, 2009); Gad Heuman and Trevor Burnard, eds., *The Routledge History of World Slavery* (London, 2011); and David Eltis and Stanley Engerman, eds., *The Cambridge World History of Slavery,* Vol. 3: *AD 1420–AD 1804* (New York, 2011).

Among the general reference works and encyclopedias on slavery, which include the American South, see especially Seymour Drescher and Stanley Engerman, eds., *A Historical Guide to World Slavery* (New York, 1997); Paul Finkelman and Joseph C. Miller, eds., *MacMillan Encyclopedia of World Slavery* (New York, 1997); Junius

P. Rodriguez, ed., *Historical Encyclopedia of World Slavery* (Santa Barbara, CA, 1997); and Junius P. Rodriguez, ed., *Slavery in the United States: A Social, Political, and Historical Encyclopedia* (Santa Barbara, CA, 2007). See also Stanley Engerman, Seymour Drescher, and Robert Paquette, *Slavery* (New York, 2001), a comprehensive collection of documents and essays.

Older key studies in the historiography of comparative slavery include Frank Tannenbaum, *Slave and Citizen: The Negro in the Americas* (New York, 1946); Herbert Klein, *Slavery in the Americas: A Comparative Study of Virginia and Cuba* (Chicago, 1967); Eugene Genovese, *The World the Slaveholders Made: Two Essays in Interpretation* (New York, 1968); Laura Foner and Eugene Genovese, eds., *Slavery in the New World: A Reader in Comparative History* (New York, 1969); Carl Degler, *Neither Black Nor White: Slavery and Race Relations in the U.S. and Brazil* (New York, 1971); Gwendolyn Midlo Hall, *Social Control in Slave Plantation Societies: A Comparison of Saint Domingue and Cuba* (Baltimore, MD, 1971); Edgar T. Thompson, *Plantation Societies, Race Relations, and the South: The Regimentation of Populations* (Durham, NC, 1975); Eugene Genovese, *From Rebellion to Revolution: Afro-American Slave Revolts in the Making of the Modern World* (Baton Rouge, LA, 1979); George Fredrickson, *White Supremacy: A Comparative Study of American and South African History* (New York, 1981); and Robin Blackburn, *The Overthrow of Colonial Slavery, 1776–1848* (London, 1988).

More recent works of particular significance include R. A. McDonald, *The Economy and Material Culture of Slaves: Goods and Chattels on the Sugar Plantations of Jamaica and Louisiana* (Baton Rouge, LA, 1993); Ira Berlin and Philip Morgan, eds., *Cultivation and Culture: Labor and the Shaping of Slave Life in the Americas* (Charlottesville, VA, 1993); Barbara Solow, ed., *Slavery and the Rise of the Atlantic System* (New York, 1993); Paul Gilroy, *The Black Atlantic: Modernity and Double Consciousness* (London, 1993); Michael Mullin, *Africa in America: Slave Acculturation and Resistance in the American South and the British Caribbean, 1736–1831* (Urbana, IL, 1995); David Barry Caspar and Darlen Clark Hine, eds., *More than Chattel: Black Women and Slavery in the Americas* (Bloomington, IN, 1996); Robin Blackburn, *The Making of New World Slavery: From the Baroque to the Modern* (London, 1997); Anthony Marx, *Making Race and Nation: A Comparison of the United States, South Africa, and Brazil* (New York, 1998); David Eltis, *The Rise of African Slavery in the Americas* (New York, 1999); Seymour Drescher, *From Slavery to Freedom: Comparative Studies in the Rise and Fall of Atlantic Slavery* (New York, 1999); Peter Kolchin, *A Sphinx on the American Land: The Nineteenth-Century South in Comparative Perspective* (Baton Rouge, LA, 2003); David Brion Davis, *Inhuman Bondage: The Rise and Fall of Slavery in the New World* (New York, 2006); Edward E. Baptist and Stephanie Camp, eds., *New Studies in the History of American Slavery* (Athens, GA, 2006); Susanna Delfino and Michelle Gillespie, eds., *Global Perspectives on Industrial Transformation in the American South* (Columbia, MO, 2006); Gwyn Campbell, Suzanne Miers, and Joseph Miller, eds., *Women and Slavery,* Vol. 2: *The Modern Atlantic* (Athens, GA, 2007); Stanley Engerman, *Slavery, Emancipation, and Freedom: Comparative Perspectives* (Baton Rouge, LA, 2007); Seymour Drescher, *Abolition: A History of Slavery and Antislavery* (New York, 2009); Robert L. Paquette and Mark M. Smith, eds., *The Oxford Handbook of Slavery in the Americas* (New York, 2010); and Robin Blackburn, *The American Crucible: Slavery, Emancipation, and Human Rights* (London, 2011).

Significant works on the recent transnational turn in American historiography and on the relationship between comparative historical studies and the transnational and world perspectives include Ian Tyrrell, *Transnational Nation: United States History in Global Perspective since 1789* (New York, 2007); Thomas Bender, *Nation Among Nations: America's Place in World History* (New York, 2006); Thomas Bender, ed., *Rethinking American History in a Global Age* (Berkeley, CA, 2002); Carl Guarneri, *America in the World: United States History in Global Context* (New York, 2006); Deborah Cohen and Maura O'Connor, eds., *Comparison and History: Europe in Cross-National Perspective* (London, 2004); Enrico Dal Lago and Constantina Katsari, eds., *Slave Systems: Ancient and Modern* (New York, 2008); and Heinz-Gerhard Haupt and Jurgen Kocka, eds., *Comparative History and the Quest for Transnationality* (Oxford, 2009). See also the older but still valid C. Van Woodward, ed., *The Comparative Approach to American History* (New York, 1968).

AMERICAN SLAVERY IN THE COLONIAL ATLANTIC

Notable recent general studies of the early modern Atlantic world include Douglas Egerton et al., *The Atlantic World: A History, 1400–1888* (New York, 2007); Toyin Falola and Kevin D. Roberts, eds., *The Atlantic World, 1450–2000* (Bloomington, IN, 2008); Thomas Benjamin, *The Atlantic World: Europeans, Africans, Indians, and Their Shared History, 1400–1900* (New York, 2009); Bernard Baylin et al., *Soundings in Atlantic History: Latent Structures and Intellectual Currents, 1500–1830* (Cambridge, MA, 2009); Jack P. Greene and Philip Morgan, eds., *Atlantic History: A Critical Appraisal* (New York, 2009); and Nicholas Canny and Philip Morgan, eds., *The Oxford Handbook of the Atlantic World, 1450–1850* (New York, 2011).

On the British Atlantic, see Eric Williams, *Capitalism and Slavery* (Chapel Hill, NC, 1944); Bernard Bailyn and Philip Morgan, eds., *Strangers within the Realm: Cultural Margins in the First British Empire* (Chapel Hill, NC, 1991); Nicholas Canny, ed., *The Oxford History of the British Empire*, Vol. I: *The Origins of Empire: British Overseas Enterprise to the Close of the Seventeenth Century* (Oxford, 1998); Peter J. Marshall, ed., *The Oxford History of the British Empire*, Vol. II: *The Eighteenth Century* (Oxford, 1998); David Armitage and Michael Braddick, eds., *The British Atlantic World, 1500–1800* (New York, 2002); and John H. Elliott, *Empires of the Atlantic World: Britain and Spain in America, 1492–1830* (New Haven, CT, 2006).

For general studies on slavery in the American South, see especially John B. Boles, *Black Southerners, 1619–1869* (Lexington, KY, 1985); Robert W. Fogel, *Without Consent or Contract: The Rise and Fall of American Slavery* (New York, 1989); Peter Kolchin, *American Slavery, 1619–1877* (New York, 2003); Ira Berlin, *Generations of Captivity: A History of African-American Slaves* (Cambridge, MA, 2004); William J. Cooper and Thomas Terrill, *The American South: A History* (New York, 2008); and Ira Berlin, *The Making of African America: The Four Great Migrations* (New York, 2010).

Notable works that stress the importance of the slaves' African origins and ethnicities include Sidney Mintz and Richard Price, *The Birth of African-American Culture: An Anthropological Perspective* (Boston, 1992, orig. pub. in 1976); Gilroy, *Black Atlantic*; John Thornton, *Africa and Africans in the Making of the Atlantic World, 1400–1800* (New York, 1998); Judith Carney, *Black Rice: The African Origins of Rice Cultivation in the Americas* (Cambridge, MA, 2001); James H. Sweet, *Recreating Africa:*

Culture, Kinship, and Religion in the Afro-Portuguese World, 1441–1700 (Chapel Hill, NC, 2003); José C. Curto and Paul Lovejoy, eds., *Enslaving Connections: Changing Cultures of Africa and Brazil during the Era of Slavery* (Amherst, NY, 2004); Gwendolyn Midlo Hall, *Slavery and African Ethnicities in the Americas: Restoring the Links* (Chapel Hill, NC, 2005); José C. Curto and Renée Soulodre-La France, eds., *Africa and the Americas: Interconnections during the Slave Trade* (Trenton, NJ, 2005); Linda Heywood and John Thornton, *Central Africans, Atlantic Creoles, and the Foundations of the Americas* (New York, 2007); James Sidbury, *Becoming African In America: Race and Nation in the Early Black Atlantic* (New York, 2009); and Fredrick C. Knight, *Working the Diaspora: The Impact of African Labor on the Anglo-American World, 1650–1850* (New York, 2010).

On the Atlantic slave trade and the middle passage in general, see Herbert S. Klein, *The Middle Passage: Comparative Studies in the Atlantic Slave Trade* (Princeton, NJ, 1978); Hugh Thomas, *The Slave Trade: A History of the Atlantic Slave Trade, 1400–1888* (London, 1999); David Eltis et al., eds., *The Trans-Atlantic Slave Trade: A Database on CD-ROM* (Cambridge, 1999); Beverly C. McMillan, ed., *Captive Passage: The Transatlantic Slave Trade and the Making of the Americas* (Washington, DC, 2002); Marcus Rediker, *The Slave Ship: A Human History* (London, 2007); Herbert S. Klein, *The Atlantic Slave Trade* (New York, 2010); and David Eltis and David Richardson, *Atlas of the Transatlantic Slave Trade* (New Haven, CT, 2010).

On the slave trades managed by different European powers, see especially Barbara L. Solow, ed., *Slavery and the Rise of the Atlantic System*; Colin Palmer, *Human Cargoes: The British Slave Trade to Spanish America, 1700–1739* (Urbana, IL, 1981); Kenneth Morgan, *Slavery, Atlantic Trade, and the British Economy, 1660–1800* (Cambridge, 2000); Luiz Felipe de Alencastro, *O trato dos viventes: Formaçao do Brasil no Atlantico Sul, seculos XVI e XVII* (Sao Paulo, 2000); M. D. D. Newitt, *A History of Portuguese Overseas Expansion, 1400–1668* (London, 2005); and Piet C. Emmer, *The Dutch Slave Trade, 1500–1850* (New York, 2006).

On the making of slavery in the Americas and its Old World background, see William Phillips, *Slavery from Roman Times to the Early Transatlantic Trade* (New York, 1985); Sidney Mintz, *Sweetness and Power: The Place of Sugar in Modern History* (New York, 1985); Philip D. Curtin, *Rise and Fall of the Plantation Complex*; Robin Blackburn, *Making of New World Slavery*; Seymour Drescher, *From Slavery to Freedom*; David Eltis, *Rise of African Slavery in the Americas*; Stuart B. Schwartz, ed., *Tropical Babylons: Sugar and the Making of the Atlantic World, 1450–1680* (Chapel Hill, NC, 2004); and Robin Blackburn, *American Crucible.*

On early slavery in Brazil, see Stuart B. Schwartz, *Sugar Plantations in the Formation of Brazilian Society: Bahia, 1550–1835* (New York, 1985); Leslie Bethell, ed., *Colonial Brazil* (New York, 1987); Stuart B. Schwartz, *Slaves, Peasants, and Rebels: Reconsidering Brazilian Slavery* (Madison, WI, 1992); and Kenneth J. Andrien, ed., *The Human Tradition in Colonial Latin America* (Wilmington, DE, 2002).

On early slavery in Barbados, see Richard S. Dunn, *Sugar and Slaves: The Rise of the Planter Class in the English West Indies, 1624–1713* (Chapel Hill, NC, 1972); Hilary M. Beckles, *White Servitude and Black Slavery in Barbados, 1627–1715* (Knoxville, TN, 1989); Russell R. Menard, *Sweet Negotiations: Sugar, Slavery, and Plantation Agriculture in Early Barbados* (Charlottesville, VA, 2006); and Susan Dwyer Amussen, *Caribbean Exchanges: Slavery and the Transformation of English Society, 1640–1700* (Chapel Hill, NC, 2007).

On early slavery in the American South, see especially Ira Berlin, *Many Thousands Gone: The First Two Centuries of Slavery in North America* (Cambridge, MA, 1998); and Michael A. Gomez, *Exchanging Our Country Marks: The Transformation of African Identities in the Colonial and Antebellum South* (Chapel Hill, NC, 1998).

On early slavery in Virginia, see Edmund Morgan, *American Slavery, American Freedom: The Ordeal of Colonial Virginia* (New York, 1975); Alan Kulikoff, *Tobacco and Slaves: The Development of Southern Cultures in the Chesapeake, 1660–1800* (Chapel Hill, NC, 1986); Alden T. Vaughan, *Roots of American Racism: Essays on the Colonial Debate* (New York, 1995); Anthony Parent, *Foul Means: The Formation of Slave Society in Virginia, 1660–1740* (Chapel Hill, NC, 2003); April Lee Hatfield, *Atlantic Virginia: Intercolonial Relations in the Seventeenth Century* (Philadelphia, PA, 2004); and Lorena S. Walsh, *Motives of Honor, Pleasure, and Profit: Plantation Management in the Colonial Chesapeake, 1607–1763* (Chapel Hill, NC, 2010).

On early slavery in South Carolina, see Peter H. Wood, *Black Majority: Negroes in Colonial South Carolina from 1670 through the Stono Rebellion* (New York, 1974); Peter Coclanis, *The Shadow of a Dream: Economic Life and Death in the South Carolina Low Country, 1660–1920* (New York, 1989); Daniel Littlefield, *Rice and Slaves: Ethnicity and the Slave Trade in Colonial South Carolina* (Urbana, IL, 1991); Jeffrey Young, *Domesticating Slavery: The Master Class in South Carolina and Georgia, 1670–1730* (Chapel Hill, NC, 1999); and S. Max Edelson, *Plantation Enterprise in Colonial South Carolina* (Cambridge, MA, 2006).

AMERICAN SLAVERY IN THE REVOLUTIONARY ATLANTIC

On eighteenth-century slavery in the American South in general, see especially Ira Berlin, *Many Thousands Gone*; Philip D. Morgan, *Slave Counterpoint: Black Life in the Eighteenth-Century Chesapeake and the Lowcountry* (Chapel Hill, NC, 1998); and Michael A. Gomez, *Exchanging Our Country Marks.*

On Virginia, see Mechal Sobel, *The World They Made Together: Black and White Values in Eighteenth-Century Virginia* (Princeton, NJ, 1989); Rhys Isaac, *The Transformation of Virginia, 1740–1790* (Chapel Hill, NC, 1989); T. H. Breen, *Tobacco Culture: The Mentality of the Great Tidewater Planters on the Eve of Revolution* (Princeton, NJ, 2001); Rhys Isaac, *Landon Carter's Uneasy Kingdom: Revolution and Rebellion on a Virginia Plantation* (New York, 2005); and Charles F. Irons, *The Origins of Proslavery Christianity: White and Black Evangelicals in Colonial and Antebellum Virginia* (Chapel Hill, NC, 2008).

On South Carolina, see especially Peter H. Wood, *Black Majority*; Daniel Littlefield, *Rice and Slaves*; Jeffrey Young, *Domesticating Slavery*; S. Max Edelson, *Plantation Enterprise in Colonial South Carolina*; Joyce Chaplin, *An Anxious Pursuit: Agricultural Innovation and Modernity in the Lower South, 1730–1815* (Chapel Hill, NC, 1993); Robert Olwell, *Masters, Slaves, and Subjects: The Culture of Power in the South Carolina Low Country, 1740–1790* (Ithaca, NY, 1998); Mark M. Smith, ed., *Stono: Documenting and Interpreting a Southern Slave Revolt* (Columbia, SC, 2006); and Peter Charles Hoffer, *Cry Liberty: The Great Stono Slave Rebellion of 1739* (New York, 2010).

On slavery in the eighteenth-century British Caribbean, see Edward Brathwaite, *The Development of Creole Society in Jamaica, 1770–1820* (Oxford, 1971); Richard B. Sheridan, *An Economic History of the British West Indies, 1623–1775* (Barbados,

1974); Mary Turner, *Slaves and Missionaries: The Disintegration of Jamaican Slave Society, 1787–1834* (Urbana, IL, 1982); Michael Craton, *Testing the Chains: Resistance to Slavery in the British West Indies* (Ithaca, NY, 1982); J. R. Ward, *British West Indian Slavery, 1750–1834: The Process of Amelioration* (Oxford, 1988); Mavis C. Campbell, *The Maroons of Jamaica, 1655–1796: A History of Resistance, Collaboration, and Betrayal* (Trenton, NJ, 1990); Arthur L. Stinchcombe, *Sugar Island Slavery in the Age of Enlightenment: The Political Economy of the Caribbean World* (Princeton, NJ, 1995); Peter J. Marshall, ed., *The Oxford History of the British Empire,* Vol. II: *The Eighteenth Century* (New York, 1998); Trevor Burnard, *Mastery, Tyranny, and Desire: Thomas Thistlewood and His Slaves in the Anglo-Jamaican World* (Chapel Hill, NC, 2004); and Vincent Brown, *The Reaper's Garden: Death and Power in the World of Atlantic Slavery* (Cambridge, MA, 2008).

On slavery in eighteenth-century Brazil, see especially Francisco Vidal Luna, *Minas Gerais: Escravos e senhores* (Sao Paulo, 1980); João José Reis and Flavio dos santos Gomes, eds., *Libertade por um fio. Historia dos Quilombos no Brasil* (Sao Paulo, 1996); Laird Bergad, *Slavery and the Demographic and Economic History of Minas Gerais, 1720–1888* (New York, 1999); Elizabeth W. Kiddy, *Blacks of the Rosary: Memory and History in Minas Gerais* (University Park, PA, 2005); Celia Maria Borges, *Escravos e libertos nas Irmandades do Rosario: Devoçao e solidaridade em Minas Gerais, seculos XVIII e XIX* (Juiz de Fora, 2005); and Herbert S. Klein and Francisco Vidal Luna, *Slavery in Brazil* (Cambridge, 2009).

On slavery and the American Revolution, see Duncan J. MacLeod, *Slavery, Race, and the American Revolution* (New York, 1974); Sylvia R. Frey, *Water from the Rock: Black Resistance in a Revolutionary Age* (Princeton, NJ, 1991); Daniel Littlefield, *Revolutionary Citizens: African Americans, 1776–1804* (New York, 1997); Paul Finkelman, *Slavery and the Founders: Race and Liberty in the Age of Jefferson* (Armonk, NY, 2001); Simon Schama, *Rough Crossings: Britain, the Slaves, and the American Revolution* (London, 2006); Cassandra Pybus, *Epic Journeys: Runaway Slaves of the American Revolution and Their Global Quest for Liberty* (Boston, 2007); and Douglas Egerton, *Death or Liberty: African Americans and Revolutionary America* (New York, 2011).

On slavery and the Haitian Revolution, see especially C. L. R. James, *The Black Jacobins: Toussaint L'Ouverture and the San Domingo Revolution* (New York, 1963, orig. pub. in 1938); Carolyn Fick, *The Making of Haiti: The Saint Domingue Revolution from Below* (Knoxville, TN, 1990); Stewart King, *Blue Coat or Powdered Wig: Free People of Color in Pre-Revolutionary Saint-Domingue* (Athens, GA, 2000); Laurent Dubois, *Avengers of the New World: The Story of the Haitian Revolution* (Cambridge, MA, 2004); John Garrigus, *Before Haiti: Race and Citizenship in French Saint-Domingue* (New York, 2006); Laurent Dubois and John D. Gerrigus, eds., *Slave Revolution in the Caribbean, 1789–1804* (New York, 2006); and Jeremy Popkin, *You Are All Free: The Haitian Revolution and the Abolition of Slavery* (New York, 2011).

On the wider context of the Haitian Revolution, see David Barry Gaspar and David Patrick Geggus, eds., *A Turbulent Time: The French Revolution and the Greater Caribbean* (Bloomington, IN, 1997); David Geggus, ed., *The Impact of the Haitian Revolution on the Atlantic World* (Columbia, SC, 2001); Gordon S. Brown, *Toussaint's Clause: The Founding Fathers and the Haitian Revolution* (Jackson, MS, 2005); Sue Peabody and Keila Grinberg, eds., *Slavery, Freedom, and the Law in the Atlantic World* (New York, 2007); Wim Klooster, *Revolutions in the Atlantic World: A Comparative History* (New York, 2009); Jane Landers, *Atlantic Creoles in the Age of*

Revolutions (Cambridge, MA, 2010); David Armitage and Sanjay Subrahmanyam, eds., *The Age of Revolutions in Global Context, c. 1760–1840* (New York, 2010); and Ashli White, *Encountering Revolution: Haiti and the Making of the Early Republic* (Baltimore, MD, 2010).

THE COTTON KINGDOM, ITS NEIGHBORS, AND ITS CONTEMPORARIES

Important general and comparative works on the second slavery and on slavery in the nineteenth-century Americas include Laura Foner and Eugene Genovese, eds., *Slavery in the New World* (Prentice Hall, NJ, 1970); Eugene Genovese, *From Rebellion to Revolution*; Ira Berlin and Philip Morgan, eds., *Cultivation and Culture*; David Barry Gaspar and Darlene Clark Hine, eds., *More than Chattel: Black Women and Slavery in the Americas* (Bloomington, IN, 1996); Dale Tomich, *Through the Prism of Slavery;* Rafael de Bivar Marquese, *Feitores do corpo, missionarios da mente. Senhores, letrados e o controle dos escravos nas Americas, 1660–1860* (Sao Paulo, 2004); Toyin Fayola and Matt D. Childs, eds., *The Yoruba Diaspora in the Atlantic World* (Bloomington, IN, 2004); Walter Johnson, *The Chattel Principle: Internal Slave Trades in North America, Brazil, and the West Indies, 1808–1888* (New Haven, CT, 2004); David Brion Davis, *Inhuman Bondage*; Laird W. Bergad, *The Comparative Histories of Slavery in Brazil, Cuba, and the United States* (New York, 2007); Jennifer L. Morgan, *Laboring Women: Reproduction and Gender in New World Slavery* (Philadelphia, PA, 2004); Gwendolyn Midlo Hall, *Slavery and African Ethnicities in the Americas*; Seymour Drescher, *Abolition*; Robert L. Paquette and Mark M. Smith, eds., *The Oxford Handbook of Slavery in the Americas*; Daniel Brett Rood, "Plantation Technocrats: A Social History of Knowledge in the Slaveholding Atlantic World, 1830–1865," Ph.D. Dissertation, University of California, Irvine, 2010; and Robin Blackburn, *The American Crucible.*

On slavery, cotton, and sugar production in the antebellum U.S. South, see Gavin Wright, *The Political Economy of the Cotton South* (New York, 1978); Mark M. Smith, *Mastered by the Clock: Time, Slavery, and Freedom in the American South* (Chapel Hill, NC, 1997); Mark M. Smith, *Debating Slavery: Economy and Society in the Antebellum American South* (New York, 1998); Peter Kolchin, *American Slavery*; Ira Berlin, *Generations of Captivity*; William K. Scarborough, *Masters of the Big House: Elite Slaveholders of the Mid-Nineteenth-Century South* (Baton Rouge, LA, 2003); Adam Rothman, *Slave Country: American Expansion and the Origins of the Deep South* (Cambridge, MA, 2005); Richard Follett, *The Sugar Masters: Planters and Slaves in Louisiana"s Cane World, 1820–1860* (Baton Rouge, LA, 2005); Susanna Delfino and Michelle Gillespie, eds., *Technology and Southern Industrialization: From the Antebellum Era to the Computer Age* (Columbia, MO, 2008); and L. Diane Barnes et al., eds., *The Old South's Modern Worlds: Slavery, Region, and Nation in the Age of Progress* (New York, 2011).

On slave culture and slave rebellion, see especially Eugene Genovese, *Roll, Jordan, Roll: The World the Slaves Made* (New York, 1974); Albert J. Rabouteau, *Slave Religion: The "Invisible Institution" in the Antebellum South* (New York, 1979); Douglas Egerton, *Gabriel's Rebellion: The Virginia Slave Conspiracies of 1800 and 1802* (Chapel Hill, NC, 1993); Brenda Stevenson, *Life in Black and White: Family and Community in the Slave South* (New York, 1996); James Sidbury, *Ploughshares into Swords: Race, Rebellion, and Identity in Gabriel's Virginia, 1730–1810* (New York, 1997);

Michael A. Gomez, *Exchanging Our Country Marks*; John Hope Franklin and Loren Schweninger, *Runaway Slaves: Rebels on the Plantation* (New York, 1999); Dylan Peningroth, *The Claims of Kinfolk: African American Property and Community in the Nineteenth-Century South* (Chapel Hill, NC, 2002); Kenneth Greenberg, ed., *Nat Turner: A Slave Rebellion in History and Memory* (New York, 2003); Douglas Egerton, *He Shall Go Free: The Lives of Denmark Vesey* (New York, 2004); Scot French, *The Rebellious Slave: Nat Turner in American Memory* (New York, 2004); and Anthony Kaye, *Joining Places: Slave Neighborhoods in the Old South* (Chapel Hill, NC, 2009).

On slavery and sugar production in Cuba, see especially Manuel Moreno Fraginals, *The Sugarmill: The Socio-Economic Complex of Sugar in Cuba, 1760–1860* (New York, 1976); Oscar Zanetti and Alejandro Garcia, *Sugar and Railroads: A Cuban History, 1837–1959* (Chapel Hill, NC, 1987); Laird Bergad, *Cuban Rural Society in the Nineteenth Century: The Social and Economic History of Monoculture in Matanzas* (Princeton, NJ, 1990); Reinaldo Funes Monzotes, *From Rainforest to Cane Field in Cuba: An Environmental History since 1492* (Chapel Hill, NC, 2008); and Jonathan Curry-Machado, *Cuban Sugar Industry: Transnational Networks and Engineering Migrants in Mid-Nineteenth-Century Cuba* (New York, 2011).

On slave resistance and slave rebellion, see Robert L. Paquette, *Sugar Is Made with Blood: The Conspiracy of La Escalera and the Conflict between Empires over Slavery in Cuba* (Princeton, NJ, 1988); Gloria Garcia Rodriguez, *Conspiraciones y Revueltas: la actividad politica de los negros en Cuba (1790–1845)*; Matt D. Childs, *The 1812 Aponte Rebellion in Cuba and the Struggle against Atlantic Slavery* (Chapel Hill, NC, 2006); and Manuel Barcia, *Seeds of Insurrection: Domination and Resistance in Western Cuban Plantations* (Baton Rouge, LA, 2008).

On slavery and coffee production in Brazil, see particularly Herbert Klein and Francisco Vidal Luna, *Slavery in Brazil* (New York, 2010); Stanley J. Stein, *Vassouras: A Brazilian Coffee County, 1850–1900* (Cambridge, MA, 1957); Warren Dean, *Rio Claro: A Brazilian Plantation System, 1820–1920* (Stanford, CA, 1976); Herbert Klein, *Slavery and the Economy of Sao Paulo, 1750–1850* (Stanford, CA, 2003); and Ricardo Salles, *E o Vale era o escravo: Vassouras, seculo XIX. Senhores e escravos no coracao do Imperio* (Rio de Janeiro, 2008).

On slave culture and slave resistance, see Roger Bastide, *The African Religions of Brazil: Toward a Sociology of the Interpretation of Civilizations* (Baltimore, MD, 2007, orig. pub. in 1960); Stuart B. Schwartz, *Slaves, Peasants, and Rebels: Reconsidering Brazilian Slavery* (Urbana, IL, 1992); João José Reis, *Slave Rebellions in Brazil: The Muslim Uprising of 1835 in Bahia* (Baltimore, MD, 1993); and João José Reis, *Death Is a Festival: Funeral Rites and Rebellion in Nineteenth-Century Brazil* (Chapel Hill, NC, 2003).

On slavery in nineteenth-century Africa, see particularly J. S. Trimingham, *Islam in West Africa* (Oxford, 1959); Susan Myers and Igor Kopytoff, eds., *Slavery in Africa: Historical and Anthropological Perspectives* (Madison, WI, 1977); Patrick Manning, *Slavery and African Life: Occidental, Oriental, and African Slave Trades* (New York, 1990); J. F. Ade Ajayi, ed., *UNESCO General History of Africa,* Vol. VI: *Africa in the Nineteenth Century until 1880* (Berkeley, CA, 1998); Paul Lovejoy, *Transformations in Slavery: A History of Slavery in Africa* (New York, 2000); and Ralph A. Austen, *Trans-Saharan Africa in World History* (New York, 2010).

On nineteenth-century slavery in the Sokoto Caliphate, see H. A. S. Johnston, *The Fulani Empire of Sokoto* (London, 1967); Murray Last, *The Sokoto Caliphate* (London,

1967); A. G. Hopkins, *An Economic History of West Africa* (New York, 1973); Sean Stilwell, *Paradoxes of Power: The Kano "Mamluks" and Male Royal Slavery in the Sokoto Caliphate, 1807–1903* (Portsmouth, NH, 2004); Paul Lovejoy, *Slavery, Commerce, and Production in the Sokoto Caliphate of West Africa* (Trenton, NJ, 2005); and Mohammed Bashir Salau, *The West African Plantation: A Case Study* (New York, 2011).

EURO-AMERICAN SERVITUDE, SHARECROPPING, AND SLAVERY

Useful comparative overviews and studies of the agrarian economy of nineteenth-century Europe and of its free and unfree systems of labor can be found in Eric J. Hobsbawm, *The Age of Capital, 1848–1875* (London, 1975); D. Spring, ed., *European Landed Elites in the Nineteenth Century* (Baltimore, MD, 1977); Jerome Blum, *The End of the Old Order in Rural Europe* (Princeton, NJ, 1978); R. Gibson and M. Blinkhorn, eds., *Landownership and Power in Modern Europe* (London, 1991); Dominic Lieven, *The Aristocracy in Europe, 1815–1914* (London, 1992); Werner Rosener, *The European Peasantries* (Oxford, 1993); Tom Scott, ed., *The Peasantries of Europe* (Harlow, 1998); Robert Gildea, *Barricades and Borders: Europe, 1800–1914* (Oxford, 2003); Hartmut Kalble, ed., *The European Way: European Societies during the Nineteenth and Twentieth Centuries* (New York, 2004); and Stefan Berger, ed., *A Companion to Nineteenth-Century Europe* (Oxford, 2009).

General and specific comparative studies on American slavery, serfdom, and systems of servitude in Europe include Peter Kolchin, *Unfree Labor: American Slavery and Russian Serfdom* (Cambridge, MA, 1987); Shearer Davis Bowman, *Masters and Lords: Mid-Nineteenth-Century U.S. Planters and Prussian Junkers* (New York, 1993); Michael L. Bush, ed., *Serfdom and Slavery: Studies in Legal Bondage* (London, 1996); Mary Turner, ed., *From Chattel Laborer to Wage Laborer* (London, 1996); Michael L. Bush, *Servitude in Modern Times* (London, 1999); Stanley Engerman, ed., *Terms of Labor: Slavery, Serfdom, and Free Labor* (Stanford, CA, 1999); Michael J. Steinfeld, *Coercion, Contract, and Labor in the Nineteenth Century* (New York, 2001); and Christopher Tomlins, *Freedom Bound: Law, Labor, and Civic Identity in Colonizing English America, 1580–1865* (New York, 2010).

Comparative studies on American slavery and Mediterranean sharecropping and tenancy, and of the American and Italian souths, include Enrico Dal Lago and Rick Halpern, eds., *The American South and the Italian Mezzogiorno: Essays in Comparative History* (New York, 2002); Don Doyle, *Nations Divided: America, Italy, and the Southern Question* (Athens, GA, 2003); and Enrico Dal Lago, *Agrarian Elites: American Slaveholders and Southern Italian Landowners, 1815–1861* (Baton Rouge, LA, 2005).

ATLANTIC ABOLITIONISM AND EUROPEAN NATIONALISM

Important comparative works on Atlantic abolitionism include Betty Fladeland, *Men and Brothers: Anglo-American Antislavery Cooperation* (Urbana, IL, 1972); David Brion Davis, *The Problem of Slavery in the Age of Revolution; Slavery and Human Progress* (New York, 1985); David Brion Davis, *Revolutions: Reflections on American Equality and Foreign Liberations* (Cambridge, MA, 1990); Richard J. Blackett, *Building an Antislavery Wall: Black Americans in the Atlantic Abolitionist Movement, 1830–1860* (Baton Rouge, LA, 1983); Robin Blackburn, *The Overthrow of Colonial Slavery*; M. P.

Kielstra, *The Politics of Slave Trade Suppression in Britain and France, 1814–1848* (London, 2000); Jeffrey Kerr-Ritchie, *Rites of August First: Emancipation Day in the Black Atlantic World* (Baton Rouge, LA, 2007); Kathryn Kish Sklar and James B. Stewart, eds., *Women's Rights and Transatlantic Slavery in the Era of Emancipation* (New Haven, CT, 2007); Seymour Drescher, *Abolition*; and Derek Peterson, ed., *Abolitionism and Imperialism in Britain, Africa, and the Atlantic* (Columbus, OH, 2010).

On American abolitionism in general and in comparative perspective, see especially James B. Stewart, *Holy Warriors: The Abolitionists and American Slavery* (New York, 1996); Paul Goodman, *Of One Blood: Abolitionism and the Origins of Racial Equality* (Berkeley, CA, 1998); Richard S. Newman, *The Transformation of American Abolitionism: Fighting Slavery in the Early Republic* (Chapel Hill, NC, 2002); Bruce Laurie, *Beyond Garrison: Antislavery and Social Reform* (New York, 2005); Timothy McCarthy et al., eds., *Prophets of Protest: Reconsidering the History of American Abolitionism* (Boston, 2006); Caleb McDaniel, "Our Country Is the World: Radical American Abolitionists Abroad," Ph.D. Dissertation, Johns Hopkins University, 2006; and Edward Bartlett Rugemer, *The Problem of Emancipation: The Caribbean Roots of the American Civil War* (Baton Rouge, LA, 2008).

On British abolitionism, see William Green, *British Slave Emancipation: The Sugar Colonies and the Great Experiment, 1830–1865* (New York, 1976); Seymour Drescher, *The Mighty Experiment: Free Labor versus Slavery in British Emancipation* (New York, 2002); Catherine Hall, *Civilising Subjects: Metropole and Colony in the English Imagination, 1830–1867* (Oxford, 2002); Christopher L. Brown, *Moral Capital: Foundations of British Abolitionism* (Chapel Hill, NC, 2006); David Ryden, *West Indian Slavery and British Abolition, 1783–1807* (New York, 2010); and Seymour Drescher, *Econocide: British Slavery in the Era of Abolition* (Chapel Hill, NC, 2010).

On Irish abolitionism, see Douglas C. Riach, "Ireland and the Campaign against American Slavery, 1830–1860," Ph.D. Dissertation, University of Edinburgh, 1975; Nini Rodgers, *Ireland, Slavery, and Anti-Slavery, 1612–1865* (New York, 2005); and Angela Murphy, *American Slavery, Irish Freedom: Abolition, Immigrant Citizenship, and the Transatlantic Movement for Irish Repeal* (Baton Rouge, LA, 2010).

On French abolitionism, see Lawrence Jennings, *French Reactions to British Emancipation* (New York, 1988); Nelly Shmidt, *Victor Schoelcher et l'abolition de l'esclavage* (Paris, 1999); and Lawrence Jennings, *French Anti-Slavery: The Movement for the Abolition of Slavery in France, 1802–1848* (New York, 2000).

On William Lloyd Garrison, see especially Henry Mayer, *All On Fire: William Lloyd Garrison and the Abolition of Slavery* (New York, 1998); and James B. Stewart, ed., *William Lloyd Garrison at Two Hundred* (New Haven, CT, 2008).

On William Henry Ashurst, see Emily Venturi, *Mr. Morris 80th Birthday Celebration, with a Brief Record of the Life of William Henry Ashurst* (London, 1904).

On Daniel O'Connell, see especially Oliver McDonagh, *O'Connell: The Life of Daniel O'Connell, 1775–1847* (Dublin, 1991); Patrick M. Geoghegan, *King Dan: The Rise of Daniel O'Connell, 1775–1829* (Dublin, 2008); and Patrick M. Geoghegan, *Liberator: The Life and Death of Daniel O'Connell, 1830–1847* (Dublin, 2010).

On Giuseppe Mazzini, see particularly William Lloyd Garrison, ed., *Joseph Mazzini: His Life, Writings, and Political Principles* (New York, 1872); Joseph Rossi, *The Image of America in Mazzini's Writings* (Madison, WI, 1954); and C. A. Bayly and Eugenio Biagini, eds., *Giuseppe Mazzini and the Globalisation of Democratic Nationalism, 1830–1920* (Oxford, 2008).

On connections and comparisons between the antebellum and Civil War United States and Europe's nineteenth-century nationalist movements, see especially Eric J. Hobsbawm, *The Age of Capital, 1848–1875* (London, 1975); Paola Gemme, *Domesticating Foreign Struggles: The Italian Risorgimento and Antebellum American Identity* (Athens, GA, 2005); and Timothy M. Roberts, *Distant Revolutions: 1848 and the Challenge to American Exceptionalism* (Charlottesville, VA, 2009).

THE AMERICAN CIVIL WAR, NATION-BUILDING, EMANCIPATION, AND ITS AFTERMATH

Important studies on the Republican Party in the context of antebellum American politics include especially Eric Foner, *Free Soil, Free Labor, Free Men: The Ideology of the Republican Party before the Civil War* (New York, 1970); William Gienapp, *The Origins of the Republican Party, 1852–1856* (New York, 1987); Don Fehrenbacher, *The Slaveholding Republic: An Account of the United States Government's Relation to Slavery* (New York, 2001); Peter J. Parish, *The North and the Nation in the Era of the Civil War* (New York, 2003); and Sean Wilentz, *The Rise of American Democracy: Jefferson to Lincoln* (New York, 2005).

On slave emancipation in the United States, see Ira Berlin et al., *Slaves No More: Three Essays on Emancipation and the Civil War* (New York, 1992); Michael Vorenberg, *Final Freedom: The Civil War, the Abolition of Slavery and the Thirteenth Amendment* (New York, 2001); Eric Foner, *Forever Free: The Story of Emancipation and Reconstruction* (New York, 2005); Harold Holzer and Sarah Vaughn Gabbard, eds., *Lincoln and Freedom: Slavery, Emancipation, and the Thirteenth Amendment* (Carbondale, IL, 2007); William A. Blair and Karen Fischer Younger, eds., *Lincoln's Proclamation: Emancipation Reconsidered* (Chapel Hill, NC, 2009); Steven Hahn, *The Political Worlds of Slavery and Freedom* (Cambridge, MA, 2009); and Stephanie McCurry, *Confederate Reckoning: Power and Politics in the Civil War South* (Cambridge, MA, 2010).

On emancipation in Cuba, see especially Rebecca J. Scott, *Slave Emancipation in Cuba: The Transition to Free Labor, 1860–1899* (Princeton, NJ, 1985); and Ada Ferrer, *Insurgent Cuba: Race, Nation, and Revolution, 1868–1898* (Chapel Hill, NC, 1999).

On emancipation in Brazil, see Leslie Bethell, *The Abolition of the Brazilian Slave Trade: Britain, Brazil, and the Slave Trade Question, 1807–1864* (Cambridge, 1970); Robert Conrad, *The Destruction of Brazilian Slavery, 1850–1888* (Berkeley, CA, 1972); and Jeffrey Needell, *The Party of Order: The Conservatives, the State, and Slavery in the Brazilian Monarchy, 1831–1871* (Stanford, CA, 2006).

On emancipation from Russian and eastern European serfdom, see David Field, *The End of Serfdom: Nobility and Bureaucracy in Russia, 1855–1861* (Cambridge, MA, 1976); Jerome Blum, *The End of the Old Order in Europe* (Princeton, NJ, 1978); Stanley L. Engerman, ed., *Terms of Labor: Slavery, Serfdom, and Free Labor* (Stanford, CA, 1999); and David Moon, *The Abolition of Serfdom in Russia, 1762–1902* (Harlow, U.K., 2001).

On nationalism in general and also on nineteenth-century Europe, see especially Eric J. Hobsbawm and Terence Ranger, eds., *The Invention of Tradition* (New York, 1983); Ernest Gellner, *Nations and Nationalism* (London, 1983); Benedict Anderson, *Imagined Communities: Reflections on the Origins and Spread of Nationalism* (London, 1983); Eric J. Hobsbawm, *Nations and Nationalism since 1780: Programme, Myth,*

Reality (New York, 1994); Patrick Lawrence, *Nationalism in Europe, 1780–1850* (London, 2004); and Timothy Bancroft and Mark Hewitson, eds., *What Is a Nation? Europe, 1789–1914* (Oxford, 2006).

On liberalism in nineteenth-century Latin America, see especially David Bushnell and Neil Macaulay, *The Emergence of Latin America in the Nineteenth Century* (New York, 1983); David Brading, *The First America: The Spanish Monarchy, Creole Patriots, and the Liberal State, 1492–1867* (New York, 1991); John Lynch, ed., *Latin American Revolutions, 1808–1826: Old and New World Regimes* (Norman, OK, 1994); and Rebecca Earle, ed., *Rumors of Wars: Civil Conflict in Nineteenth-Century Latin America* (London, 2000).

On the American Civil War, nation-building, and nationalism in comparative perspective, see Carl Degler, *One Among Many: The Civil War in Comparative Perspective* (New York, 1992); Liah Greenfeld, *Nationalism: Five Roads to Modernity* (Cambridge, MA, 1995); Stig Forster and Jörg Nagler, eds., *On the Road to Total War: The American Civil War and the German Wars of Unification, 1861–1871* (New York, 1997); Brian Holden Reid, *The American Civil War and the Wars of the Industrial Revolution* (London, 1999); James McPherson, *Is Blood Thicker than Water? Crises of Nationalism in the Modern World* (New York, 1999); Nicholas and Peter Onuf, *Nations, Markets, and Wars: Modern History and the American Civil War* (Charlottesville, VA, 2006); James Belich, *Replenish the Earth*; and Amanda Foreman, *A World on Fire: An Epic History of Two Nations Divided* (New York, 2010).

On nationalism and nation-building in the United States, North and South, see especially Drew Faust, *The Creation of Confederate Nationalism: Ideology and Identity in the Civil War South* (Baton Rouge, LA, 1988); Susan-Mary Grant, *North over South: Northern Nationalism and America Identity in the Antebellum Era* (Lawrence, KS, 2000); Melinda Lawson, *Patriot Fires: Forging a New American Nationalism in the Civil War North* (Lawrence, KS, 2002); Ann Sarah Rubin, *A Shattered Nation: The Rise and Fall of the Confederacy, 1816–1868* (Chapel Hill, NC, 2007); and Susan-Mary Grant, *The War for a Nation: The American Civil War* (New York, 2006).

On Abraham Lincoln and the reforging of the American nation, see especially Gary Wills, *Lincoln at Gettysburg: The Words that Remade America* (New York, 1992); Philip S. Paludan, *The Presidency of Abraham Lincoln* (Lawrence, KS, 1994); David Herbert Donald, *Lincoln* (New York, 1995); Howard Belz, *Abraham Lincoln, Constitutionalism, and Equal Rights in the Civil War Era* (New York, 1998); Harry V. Jaffa, *A New Birth of Freedom: Abraham Lincoln and the Coming of the Civil War* (New York, 2000); Richard Carwardine, *Abraham Lincoln: A Life of Purpose and Power* (New York, 2005); and Eric Foner, *A Fiery Trial: Abraham Lincoln and American Slavery* (New York, 2010).

On the Italian *Risorgimento,* see especially Denis Mack Smith, ed., *The Making of Italy* (London, 1986); Alberto Mario Banti, *Il Risorgimento italiano* (Rome-Bari, 2007); Mario Isnenghi and Eva Cecchinato, eds., *Fare l'Italia: Unità e disunità nel Risorgimento* (Turin, 2008); and Lucy Riall, *Risorgimento: The History of Italy from Napoleon to Nation State* (New York, 2009).

On Count Camillo Cavour and Italy's Moderate Liberals, see especially Rosario Romeo, *Dal Piemonte sabaudo all'Italia liberale* (Rome-Bari, 1963); Rosario Romeo, *Vita di Cavour* (Rome-Bari, 1985); Luciano Cafagna, *Cavour* (Bologna, 1999); and Adriano Viarengo, *Cavour* (Salerno, 2010).

For comparisons between post-emancipation societies in the Americas, see especially Eric Foner, *Nothing but Freedom: Emancipation and Its Legacy* (Baton Rouge, LA, 1983); Frederick Cooper et al., *Beyond Slavery: Explorations of Race, Labor, and Citizenship in Postemancipation Societies* (Chapel Hill, NC, 2000); Rebecca Scott, *Degrees of Freedom: Louisiana and Cuba after Slavery* (Baton Rouge, LA, 2005); Seymour Drescher, *Abolition*; Robert L. Paquette and Mark M. Smith, eds., *Oxford Handbook of Slavery*; and Robin Blackburn, *American Crucible.*

For comparisons with Europe, see especially Michael L. Bush, ed., *Serfdom and Slavery*; Stanley Engerman, ed., *Terms of Labor*; Michael L. Bush, *Servitude in Modern Times*; and Stefan Berger, *Nineteenth Century Europe.*

On emancipation in Africa, see J. F. Ade Ajay, ed., *Africa in the Nineteenth Century*; A. Adu Boahen, ed., *UNESCO General History of Africa*, Vol. VII: *Africa under Colonial Domination, 1880–1935* (Berkeley, CA, 1990); and Paul Lovejoy, *Transformations in Slavery.*

INDEX

ABOUT THE AUTHOR

Enrico Dal Lago is Lecturer in American History at the National University of Ireland, Galway, where he teaches both American slavery and comparative slavery. He is the author of *Agrarian Elites: American Slaveholders and Southern Italian Landowners, 1815–1861* (Baton Rouge: Louisiana State University Press, 2005). He is also the coeditor of (with Rick Halpern) and a contributor to *The American South and the Italian Mezzogiorno: Essays in Comparative History* (New York: Palgrave, 2001), *Slavery and Emancipation* (Oxford: Blackwell, 2002), and (with Constantina Katsari) *Slave Systems: Ancient and Modern* (New York: Cambridge University Press, 2008). He is currently working on a comparative history of American abolitionism and Italian democratic nationalism focusing on William Lloyd Garrison and Giuseppe Mazzini.